FREE Test Taking Tips DVD Offer

To help us better serve you, we have developed a Test Taking Tips DVD that we would like to give you for FREE. **This DVD covers world-class test taking tips that you can use to be even more successful when you are taking your test.**

All that we ask is that you email us your feedback about your study guide. Please let us know what you thought about it – whether that is good, bad or indifferent.

To get your **FREE Test Taking Tips DVD**, email freedvd@studyguideteam.com with "FREE DVD" in the subject line and the following information in the body of the email:

  a. The title of your study guide.

  b. Your product rating on a scale of 1-5, with 5 being the highest rating.

  c. Your feedback about the study guide. What did you think of it?

  d. Your full name and shipping address to send your free DVD.

If you have any questions or concerns, please don't hesitate to contact us at freedvd@studyguideteam.com.

Thanks again!

SAT Prep Questions Book 2021-2022

Book 2021-2022

3 SAT Practice Tests with Detailed Answer Explanations for the College Board Exam [3rd Edition]

Joshua Rueda

Interested in buying more than 10 copies of our product? Contact us about bulk discounts:
bulkorders@studyguideteam.com

ISBN 13: 9781637756898
ISBN 10: 1637756895

Table of Contents

//Test Prep Books!!!

Quick Overview

As you draw closer to taking your exam, effective preparation becomes more and more important. Thankfully, you have this study guide to help you get ready. Use this guide to help keep your studying on track and refer to it often.

This study guide contains several key sections that will help you be successful on your exam. The guide contains tips for what you should do the night before and the day of the test. Also included are test-taking tips. Knowing the right information is not always enough. Many well-prepared test takers struggle with exams. These tips will help equip you to accurately read, assess, and answer test questions.

A large part of the guide is devoted to showing you what content to expect on the exam and to helping you better understand that content. In this guide are practice test questions so that you can see how well you have grasped the content. Then, answer explanations are provided so that you can understand why you missed certain questions.

Don't try to cram the night before you take your exam. This is not a wise strategy for a few reasons. First, your retention of the information will be low. Your time would be better used by reviewing information you already know rather than trying to learn a lot of new information. Second, you will likely become stressed as you try to gain a large amount of knowledge in a short amount of time. Third, you will be depriving yourself of sleep. So be sure to go to bed at a reasonable time the night before. Being well-rested helps you focus and remain calm.

Be sure to eat a substantial breakfast the morning of the exam. If you are taking the exam in the afternoon, be sure to have a good lunch as well. Being hungry is distracting and can make it difficult to focus. You have hopefully spent lots of time preparing for the exam. Don't let an empty stomach get in the way of success!

When travelling to the testing center, leave earlier than needed. That way, you have a buffer in case you experience any delays. This will help you remain calm and will keep you from missing your appointment time at the testing center.

Be sure to pace yourself during the exam. Don't try to rush through the exam. There is no need to risk performing poorly on the exam just so you can leave the testing center early. Allow yourself to use all of the allotted time if needed.

Remain positive while taking the exam even if you feel like you are performing poorly. Thinking about the content you should have mastered will not help you perform better on the exam.

Once the exam is complete, take some time to relax. Even if you feel that you need to take the exam again, you will be well served by some down time before you begin studying again. It's often easier to convince yourself to study if you know that it will come with a reward!

Test-Taking Strategies

1. Predicting the Answer

When you feel confident in your preparation for a multiple-choice test, try predicting the answer before reading the answer choices. This is especially useful on questions that test objective factual knowledge. By predicting the answer before reading the available choices, you eliminate the possibility that you will be distracted or led astray by an incorrect answer choice. You will feel more confident in your selection if you read the question, predict the answer, and then find your prediction among the answer choices. After using this strategy, be sure to still read all of the answer choices carefully and completely. If you feel unprepared, you should not attempt to predict the answers. This would be a waste of time and an opportunity for your mind to wander in the wrong direction.

2. Reading the Whole Question

Too often, test takers scan a multiple-choice question, recognize a few familiar words, and immediately jump to the answer choices. Test authors are aware of this common impatience, and they will sometimes prey upon it. For instance, a test author might subtly turn the question into a negative, or he or she might redirect the focus of the question right at the end. The only way to avoid falling into these traps is to read the entirety of the question carefully before reading the answer choices.

3. Looking for Wrong Answers

Long and complicated multiple-choice questions can be intimidating. One way to simplify a difficult multiple-choice question is to eliminate all of the answer choices that are clearly wrong. In most sets of answers, there will be at least one selection that can be dismissed right away. If the test is administered on paper, the test taker could draw a line through it to indicate that it may be ignored; otherwise, the test taker will have to perform this operation mentally or on scratch paper. In either case, once the obviously incorrect answers have been eliminated, the remaining choices may be considered. Sometimes identifying the clearly wrong answers will give the test taker some information about the correct answer. For instance, if one of the remaining answer choices is a direct opposite of one of the eliminated answer choices, it may well be the correct answer. The opposite of obviously wrong is obviously right! Of course, this is not always the case. Some answers are obviously incorrect simply because they are irrelevant to the question being asked. Still, identifying and eliminating some incorrect answer choices is a good way to simplify a multiple-choice question.

4. Don't Overanalyze

Anxious test takers often overanalyze questions. When you are nervous, your brain will often run wild, causing you to make associations and discover clues that don't actually exist. If you feel that this may be a problem for you, do whatever you can to slow down during the test. Try taking a deep breath or counting to ten. As you read and consider the question, restrict yourself to the particular words used by the author. Avoid thought tangents about what the author *really* meant, or what he or she was *trying* to say. The only things that matter on a multiple-choice test are the words that are actually in the question. You must avoid reading too much into a multiple-choice question, or supposing that the writer meant something other than what he or she wrote.

5. No Need for Panic

It is wise to learn as many strategies as possible before taking a multiple-choice test, but it is likely that you will come across a few questions for which you simply don't know the answer. In this situation, avoid panicking. Because most multiple-choice tests include dozens of questions, the relative value of a single wrong answer is small. As much as possible, you should compartmentalize each question on a multiple-choice test. In other words, you should not allow your feelings about one question to affect your success on the others. When you find a question that you either don't understand or don't know how to answer, just take a deep breath and do your best. Read the entire question slowly and carefully. Try rephrasing the question a couple of different ways. Then, read all of the answer choices carefully. After eliminating obviously wrong answers, make a selection and move on to the next question.

6. Confusing Answer Choices

When working on a difficult multiple-choice question, there may be a tendency to focus on the answer choices that are the easiest to understand. Many people, whether consciously or not, gravitate to the answer choices that require the least concentration, knowledge, and memory. This is a mistake. When you come across an answer choice that is confusing, you should give it extra attention. A question might be confusing because you do not know the subject matter to which it refers. If this is the case, don't eliminate the answer before you have affirmatively settled on another. When you come across an answer choice of this type, set it aside as you look at the remaining choices. If you can confidently assert that one of the other choices is correct, you can leave the confusing answer aside. Otherwise, you will need to take a moment to try to better understand the confusing answer choice. Rephrasing is one way to tease out the sense of a confusing answer choice.

7. Your First Instinct

Many people struggle with multiple-choice tests because they overthink the questions. If you have studied sufficiently for the test, you should be prepared to trust your first instinct once you have carefully and completely read the question and all of the answer choices. There is a great deal of research suggesting that the mind can come to the correct conclusion very quickly once it has obtained all of the relevant information. At times, it may seem to you as if your intuition is working faster even than your reasoning mind. This may in fact be true. The knowledge you obtain while studying may be retrieved from your subconscious before you have a chance to work out the associations that support it. Verify your instinct by working out the reasons that it should be trusted.

8. Key Words

Many test takers struggle with multiple-choice questions because they have poor reading comprehension skills. Quickly reading and understanding a multiple-choice question requires a mixture of skill and experience. To help with this, try jotting down a few key words and phrases on a piece of scrap paper. Doing this concentrates the process of reading and forces the mind to weigh the relative importance of the question's parts. In selecting words and phrases to write down, the test taker thinks about the question more deeply and carefully. This is especially true for multiple-choice questions that are preceded by a long prompt.

9. Subtle Negatives

One of the oldest tricks in the multiple-choice test writer's book is to subtly reverse the meaning of a question with a word like *not* or *except*. If you are not paying attention to each word in the question, you can easily be led astray by this trick. For instance, a common question format is, "Which of the following is…?" Obviously, if the question instead is, "Which of the following is not…?," then the answer will be quite different. Even worse, the test makers are aware of the potential for this mistake and will include one answer choice that would be correct if the question were not negated or reversed. A test taker who misses the reversal will find what he or she believes to be a correct answer and will be so confident that he or she will fail to reread the question and discover the original error. The only way to avoid this is to practice a wide variety of multiple-choice questions and to pay close attention to each and every word.

10. Reading Every Answer Choice

It may seem obvious, but you should always read every one of the answer choices! Too many test takers fall into the habit of scanning the question and assuming that they understand the question because they recognize a few key words. From there, they pick the first answer choice that answers the question they believe they have read. Test takers who read all of the answer choices might discover that one of the latter answer choices is actually *more* correct. Moreover, reading all of the answer choices can remind you of facts related to the question that can help you arrive at the correct answer. Sometimes, a misstatement or incorrect detail in one of the latter answer choices will trigger your memory of the subject and will enable you to find the right answer. Failing to read all of the answer choices is like not reading all of the items on a restaurant menu: you might miss out on the perfect choice.

11. Spot the Hedges

One of the keys to success on multiple-choice tests is paying close attention to every word. This is never truer than with words like almost, most, some, and sometimes. These words are called "hedges" because they indicate that a statement is not totally true or not true in every place and time. An absolute statement will contain no hedges, but in many subjects, the answers are not always straightforward or absolute. There are always exceptions to the rules in these subjects. For this reason, you should favor those multiple-choice questions that contain hedging language. The presence of qualifying words indicates that the author is taking special care with his or her words, which is certainly important when composing the right answer. After all, there are many ways to be wrong, but there is only one way to be right! For this reason, it is wise to avoid answers that are absolute when taking a multiple-choice test. An absolute answer is one that says things are either all one way or all another. They often include words like *every*, *always*, *best*, and *never*. If you are taking a multiple-choice test in a subject that doesn't lend itself to absolute answers, be on your guard if you see any of these words.

12. Long Answers

In many subject areas, the answers are not simple. As already mentioned, the right answer often requires hedges. Another common feature of the answers to a complex or subjective question are qualifying clauses, which are groups of words that subtly modify the meaning of the sentence. If the question or answer choice describes a rule to which there are exceptions or the subject matter is complicated, ambiguous, or confusing, the correct answer will require many words in order to be expressed clearly and accurately. In essence, you should not be deterred by answer choices that seem excessively long. Oftentimes, the author of the text will not be able to write the correct answer without offering some qualifications and modifications. Your job is to read the answer choices thoroughly and completely and to select the one that most accurately and precisely answers the question.

13. Restating to Understand

Sometimes, a question on a multiple-choice test is difficult not because of what it asks but because of how it is written. If this is the case, restate the question or answer choice in different words. This process serves a couple of important purposes. First, it forces you to concentrate on the core of the question. In order to rephrase the question accurately, you have to understand it well. Rephrasing the question will concentrate your mind on the key words and ideas. Second, it will present the information to your mind in a fresh way. This process may trigger your memory and render some useful scrap of information picked up while studying.

14. True Statements

Sometimes an answer choice will be true in itself, but it does not answer the question. This is one of the main reasons why it is essential to read the question carefully and completely before proceeding to the answer choices. Too often, test takers skip ahead to the answer choices and look for true statements. Having found one of these, they are content to select it without reference to the question above. Obviously, this provides an easy way for test makers to play tricks. The savvy test taker will always read the entire question before turning to the answer choices. Then, having settled on a correct answer choice, he or she will refer to the original question and ensure that the selected answer is relevant. The mistake of choosing a correct-but-irrelevant answer choice is especially common on questions related to specific pieces of objective knowledge. A prepared test taker will have a wealth of factual knowledge at his or her disposal, and should not be careless in its application.

15. No Patterns

One of the more dangerous ideas that circulates about multiple-choice tests is that the correct answers tend to fall into patterns. These erroneous ideas range from a belief that B and C are the most common right answers, to the idea that an unprepared test-taker should answer "A-B-A-C-A-D-A-B-A." It cannot be emphasized enough that pattern-seeking of this type is exactly the WRONG way to approach a multiple-choice test. To begin with, it is highly unlikely that the test maker will plot the correct answers according to some predetermined pattern. The questions are scrambled and delivered in a random order. Furthermore, even if the test maker was following a pattern in the assignation of correct answers, there is no reason why the test taker would know which pattern he or she was using. Any attempt to discern a pattern in the answer choices is a waste of time and a distraction from the real work of taking the test. A test taker would be much better served by extra preparation before the test than by reliance on a pattern in the answers.

FREE DVD OFFER

Don't forget that doing well on your exam includes both understanding the test content and understanding how to use what you know to do well on the test. We offer a completely FREE Test Taking Tips DVD that covers world class test taking tips that you can use to be even more successful when you are taking your test.

All that we ask is that you email us your feedback about your study guide. To get your **FREE Test Taking Tips DVD**, email freedvd@studyguideteam.com with "FREE DVD" in the subject line and the following information in the body of the email:

- The title of your study guide.
- Your product rating on a scale of 1-5, with 5 being the highest rating.
- Your feedback about the study guide. What did you think of it?
- Your full name and shipping address to send your free DVD.

Introduction to the SAT

Function of the Test

The SAT is a standardized test taken by high school students across the United States and given internationally for college placement. It is designed to measure problem solving ability, communication, and understanding complex relationships. The SAT also serves as a qualifying measure to identify students for college scholarships, depending on the college being applied to. All colleges in the U.S. accept the SAT, and, in addition to admissions and scholarships, use SAT scores for course placement as well as academic counseling.

Most of the high school students who take the SAT are seniors. In 2016, the number of students who took the SAT was just under 1.7 million. It's important to note that since many updates have been implemented during the 2016 year, the data points cannot be compared to those in previous years. In 2014, 42.6 percent of students met the College Board's "college and career readiness" benchmark, and in 2015, 41.9 percent met this benchmark.

Test Administration

The SAT is offered on seven days throughout the year at schools throughout the United States. Internationally, the SAT is offered on five days throughout the year. There are thousands of testing centers worldwide. Test-takers can view the test centers in their area when they register for the test, or they can view testing locations at the College Board website, a not-for-profit that owns and publishes the SAT.

The SAT registration fee is $46, and the SAT with Essay registration fee is $60, although both of these have fee waivers available. Also note that students outside the U.S. may have to pay an extra processing fee. Additional fees include registering by phone, changing fee, late registration fee, or a waitlist fee. Test-takers may register four score reports for free up to nine days after the test. Any additional score reports cost $12, although fee waivers are available for this as well.

Test Format

The SAT gauges a student's proficiency in three areas: Reading, Mathematics, and Writing and Language. The reading portion of the SAT measures comprehension, requiring candidates to read multi-paragraph fiction and non-fiction segments including informational visuals, such as charts, tables and graphs, and answer questions based on this content. Fluency in problem solving, conceptual understanding of equations, and real-world applications are characteristics of the math test. The writing and language portion requires students to evaluate and edit writing and graphics to obtain an answer that correctly conveys the information given in the passage.

The SAT contains 154 multiple-choice questions, with each section comprising over 40 questions. A different length of time is given for each section, for a total of three hours, plus fifty minutes for the essay (optional).

Section	Time (In Minutes)	Number of Questions
Reading	65	52
Writing and Language	35	44
Mathematics	80	58
Essay (optional)	50 (optional)	1 (optional)
Total	**180**	**154 + optional essay**

Scoring

Scores for the new SAT are based on a scale from 400 to 1600. Scores range from 200 to 800 for Evidence-Based Reading and Writing, and 200 to 800 for Math. The optional essay is scored from 2 to 8. The SAT also no longer penalizes for incorrect answers. Therefore, a student's raw score is the number of correctly answered questions.

On the College Board website, there are indicators to determine what the benchmark scores are. The scores are divided up into green, yellow, or red. Green meets or exceeds the benchmark, and shows a 480 to 800 in Evidence-Based Reading and Writing, and a 530 to 800 in Math.

Recent/Future Developments

Starting June 2021, the optional SAT essay will be discontinued in most states. In schools where the students take the SAT as part of SAT School Day administrations, the optional SAT may still be available, but students should check with their school first. For students who take the SAT on a Saturday, the optional essay will not be available.

SAT Practice Test #1

Reading Test

Fiction

Questions 1–10 are based on the following passage:

We made it. We created it. We brought it forth from the night of the ages. We alone. Our hands. Our mind. Ours alone and only.

We know not what we are saying. Our head is reeling. We look upon the light which we have made. We shall be forgiven for anything we say tonight...

Tonight, after more days and trials than we can count, we finished building a strange thing, from the remains of the Unmentionable Times, a box of glass, devised to give forth the power of the sky of greater strength than we had ever achieved before. And when we put our wires to this box, when we closed the current—the wire glowed! It came to life, it turned red, and a circle of light lay on the stone before us.

We stood, and we held our head in our hands. We could not conceive of that which we had created. We had touched no flint, made no fire. Yet here was light, light that came from nowhere, light from the heart of metal.

We blew out the candle. Darkness swallowed us. There was nothing left around us, nothing save night and a thin thread of flame in it, as a crack in the wall of a prison. We stretched our hands to the wire, and we saw our fingers in the red glow. We could not see our body nor feel it, and in that moment nothing existed save our two hands over a wire glowing in a black abyss.

Then we thought of the meaning of that which lay before us. We can light our tunnel, and the City, and all the Cities of the world with nothing save metal and wires. We can give our brothers a new light, cleaner and brighter than any they have ever known. The power of the sky can be made to do men's bidding. There are no limits to its secrets and its might, and it can be made to grant us anything if we but choose to ask.

Then we knew what we must do. Our discovery is too great for us to waste our time in sweeping the streets. We must not keep our secret to ourselves, nor buried under the ground. We must bring it into the sight of all men. We need all our time, we need the work rooms of the Home of the Scholars, we want the help of our brother Scholars and their wisdom joined to ours. There is so much work ahead for all of us, for all the Scholars of the world.

In a month, the World Council of Scholars is to meet in our City. It is a great Council, to which the wisest of all lands are elected, and it meets once a year in the different Cities of the earth. We shall go to this Council and we shall lay before them, as our gift, this glass box with the power of the sky. We shall confess everything to them. They will see, understand and forgive. For our gift is greater than our transgression. They will explain it to the Council of Vocations, and we shall be assigned to the Home of the Scholars. This has never been done before, but neither has a gift such as ours ever been offered to men.

We must wait. We must guard our tunnel as we had never guarded it before. For should any men save the Scholars learn of our secret, they would not understand it, nor would they believe us. They would see nothing, save our crime of working alone, and they would destroy us and our light. We care not about our body, but our light is...

Yes, we do care. For the first time do we care about our body. For this wire is as a part of our body, as a vein torn from us, glowing with our blood. Are we proud of this thread of metal, or of our hands which made it, or is there a line to divide these two?

Excerpt from *Anthem* by Ayn Rand

1. What is the overall tone of this passage?
   a. Dreary
   b. Unnerving
   c. Excited
   d. Humorous

2. Why must the invention be kept a secret?
   a. It is illegal to work alone.
   b. The Home of the Scholars will try to take credit.
   c. They were supposed to be sweeping streets.
   d. Remains from the Unmentionable Times are off limits.

3. Which literary device is used in the following sentence from paragraph five?

   Darkness swallowed us.

   a. Metaphor
   b. Synecdoche
   c. Flashback
   d. Personification

4. What does the narrator compare their discovery to?
   a. The sun
   b. A candle
   c. Prison
   d. Blood

5. Why does the narrator expect to be forgiven?
   a. Their invention will save the City money.
   b. The possibilities of the invention outweigh their crime.
   c. They will apologize to the World Council of Scholars.
   d. Their invention is part of their body.

6. What is the meaning of the word *transgression* mean in paragraph eight?
   a. Obedience
   b. Disruption
   c. Confession
   d. Offense

7. Which quote is an example of personification?
    a. "Our head is reeling."
    b. "Darkness swallowed us."
    c. "We stood, and we held our head in our hands."
    d. "We care not about our body, but our light is…"

8. What will NOT help advance the narrator's discovery?
    a. Using the work rooms of the Home of the Scholars
    b. Gaining wisdom from the brother Scholars
    c. Keeping the discovery a secret from the World Council of Scholars
    d. Stopping the sweeping of the streets

9. How does the narrator feel their status will change as a result of their invention?
    a. They will advance from street sweeper to the Home of the Scholars.
    b. Their status will remain the same after their confession.
    c. They will rise in status from tunnel worker to the Council of Vocations.
    d. They are not sure how their status will change due to the crime they committed.

10. Which transformation has happened by the end of the passage?
    a. The narrator has been forgiven for their transgression.
    b. The gift of electricity has been shared with all the cities of the world.
    c. The discovery has made it easier to sweep streets and light tunnels.
    d. The narrator recognizes the significance of their body.

History/Social Studies

Questions 11–20 are based upon the following passage:

Our people are losing that faith, not only in government itself but in the ability as citizens to serve as the ultimate rulers and shapers of our democracy. As a people we know our past and we are proud of it. Our progress has been part of the living history of America, even the world. We always believed that we were part of a great movement of humanity itself called democracy, involved in the search for freedom, and that belief has always strengthened us in our purpose. But just as we are losing our confidence in the future, we are also beginning to close the door on our past.

In a nation that was proud of hard work, strong families, close-knit communities, and our faith in God, too many of us now tend to worship self-indulgence and consumption. Human identity is no longer defined by what one does, but by what one owns. But we've discovered that owning things and consuming things does not satisfy our longing for meaning. We've learned that piling up material goods cannot fill the emptiness of lives which have no confidence or purpose.

The symptoms of this crisis of the American spirit are all around us. For the first time in the history of our country a majority of our people believe that the next five years will be worse than the past five years. Two-thirds of our people do not even vote. The productivity of American workers is actually dropping, and the willingness of Americans to save for the future has fallen below that of all other people in the Western world.

As you know, there is a growing disrespect for government and for churches and for schools, the news media, and other institutions. This is not a message of happiness or reassurance, but it is the truth and it is a warning.

These changes did not happen overnight. They've come upon us gradually over the last generation, years that were filled with shocks and tragedy.

We were sure that ours was a nation of the ballot, not the bullet, until the murders of John Kennedy and Robert Kennedy and Martin Luther King, Jr. We were taught that our armies were always invincible and our causes were always just, only to suffer the agony of Vietnam. We respected the Presidency as a place of honor until the shock of Watergate.

We remember when the phrase "sound as a dollar" was an expression of absolute dependability, until ten years of inflation began to shrink our dollar and our savings. We believed that our Nation's resources were limitless until 1973, when we had to face a growing dependence on foreign oil.

These wounds are still very deep. They have never been healed. Looking for a way out of this crisis, our people have turned to the Federal Government and found it isolated from the mainstream of our Nation's life. Washington, D.C., has become an island. The gap between our citizens and our Government has never been so wide. The people are looking for honest answers, not easy answers; clear leadership, not false claims and evasiveness and politics as usual.

What you see too often in Washington and elsewhere around the country is a system of government that seems incapable of action. You see a Congress twisted and pulled in every direction by hundreds of well-financed and powerful special interests. You see every extreme position defended to the last vote, almost to the last breath by one unyielding group or another. You often see a balanced and a fair approach that demands sacrifice, a little sacrifice from everyone, abandoned like an orphan without support and without friends.

Often you see paralysis and stagnation and drift. You don't like it, and neither do I. What can we do?

First of all, we must face the truth, and then we can change our course. We simply must have faith in each other, faith in our ability to govern ourselves, and faith in the future of this Nation. Restoring that faith and that confidence to America is now the most important task we face. It is a true challenge of this generation of Americans.

Excerpt from "The Crisis of Confidence" by Jimmy Carter

11. What is the underlying message of Jimmy Carter's speech?
    a. There is no hope for the future of the United States.
    b. The American people have lost faith in their government.
    c. Finding the way again as a nation will be hard and will require facing the truth.
    d. America is not as great as other Western countries.

12. What is NOT a symptom of the crisis of the American spirit?
    a. Too many people tend to worship self-indulgence and consumption.
    b. Respect for government, churches and schools is growing.
    c. Two-thirds of American people do not vote.
    d. The willingness to save for the future has fallen.

13. Which phrase could replace "sound as a dollar" in paragraph seven?
    a. Hands down
    b. Solid as a rock
    c. Piece of cake
    d. Fair and square

14. Pointing out the shortcomings of government makes President Carter sound like he is:
    a. Unreliable as a leader
    b. Desperate for acceptance
    c. Relatable and empathetic
    d. Condescending toward citizens

15. What does Carter mean, metaphorically, when he says that "Washington, D.C. has become an island"?
    a. Members of Congress are relaxing on vacation while the people suffer.
    b. The government is using limited resources unwisely.
    c. The White House is an oasis for Americans in need.
    d. The government has isolated itself from its citizens.

16. What is the purpose of paragraphs six and seven?
    a. To point out that previous presidents have made mistakes
    b. To provide examples of why people are losing respect for government and other institutions
    c. To prove that our past is full of tragedy and our future is full of hope
    d. To suggest Americans' expectations are too high

17. According to President Carter, what is "politics as usual"?
    a. Decisions that do not benefit the executive branch
    b. Jargon that does not make sense to the average American
    c. A way of proceeding that is full of false claims, evasiveness, and inactivity
    d. A balanced and fair approach that demands sacrifice

18. In the second-to-last paragraph, the word *stagnation* means a lack of:
    a. Progress
    b. Understanding
    c. Knowledge
    d. Empathy

19. How does the tone shift from the beginning to the end of the passage?
    a. The tone starts informally but ends seriously.
    b. The tone starts elevated but ends unremarkably.
    c. The tone starts bleakly but ends optimistically.
    d. The tone starts preachy but ends humbly.

20. What is Carter's proposed solution to the crisis of confidence in America?
    a. Healing the wounds, shocks, and tragedies of the past
    b. Removing special interest groups from the government
    c. Restoring faith in ourselves, democracy, and our nation
    d. Reverting back to strong families and close-knit communities

History/Social Studies

Questions 21–30 is based on the following passage:

The peoples of a number of countries of the world have recently had totalitarian regimes forced upon them against their will. The Government of the United States has made frequent protests against coercion and intimidation, in violation of the Yalta agreement, in Poland, Rumania, and Bulgaria. I must also state that in a number of other countries there have been similar developments.

At the present moment in world history nearly every nation must choose between alternative ways of life. The choice is too often not a free one.

One way of life is based upon the will of the majority, and is distinguished by free institutions, representative government, free elections, guarantees of individual liberty, freedom of speech and religion, and freedom from political oppression.

The second way of life is based upon the will of a minority forcibly imposed upon the majority. It relies upon terror and oppression, a controlled press and radio; fixed elections, and the suppression of personal freedoms.

I believe that it must be the policy of the United States to support free peoples who are resisting attempted subjugation by armed minorities or by outside pressures.

I believe that we must assist free peoples to work out their own destinies in their own way.

I believe that our help should be primarily through economic and financial aid which is essential to economic stability and orderly political processes.

The world is not static, and the status quo is not sacred. But we cannot allow changes in the status quo in violation of the Charter of the United Nations by such methods as coercion, or by such subterfuges as political infiltration. In helping free and independent nations to maintain their freedom, the United States will be giving effect to the principles of the Charter of the United Nations.

It is necessary only to glance at a map to realize that the survival and integrity of the Greek nation are of grave importance in a much wider situation. If Greece should fall under the control of an armed minority, the effect upon its neighbor, Turkey, would be immediate and serious. Confusion and disorder might well spread throughout the entire Middle East.

Moreover, the disappearance of Greece as an independent state would have a profound effect upon those countries in Europe whose peoples are struggling against great difficulties to maintain their freedoms and their independence while they repair the damages of war.

It would be an unspeakable tragedy if these countries, which have struggled so long against overwhelming odds, should lose that victory for which they sacrificed so much. Collapse of free institutions and loss of independence would be disastrous not only for them but for the world. Discouragement and possibly failure would quickly be the lot of neighboring peoples striving to maintain their freedom and independence.

Should we fail to aid Greece and Turkey in this fateful hour, the effect will be far reaching to the West as well as to the East.

We must take immediate and resolute action.

I therefore ask the Congress to provide authority for assistance to Greece and Turkey in the amount of $400,000,000 for the period ending June 30, 1948. In requesting these funds, I have taken into consideration the maximum amount of relief assistance which would be furnished to Greece out of the $350,000,000 which I recently requested that the Congress authorize for the prevention of starvation and suffering in countries devastated by the war.

In addition to funds, I ask the Congress to authorize the detail of American civilian and military personnel to Greece and Turkey, at the request of those countries, to assist in the tasks of reconstruction, and for the purpose of supervising the use of such financial and material assistance as may be furnished. I recommend that authority also be provided for the instruction and training of selected Greek and Turkish personnel.

Finally, I ask that the Congress provide authority which will permit the speediest and most effective use, in terms of needed commodities, supplies, and equipment, of such funds as may be authorized.

If further funds, or further authority, should be needed for purposes indicated in this message, I shall not hesitate to bring the situation before the Congress. On this subject the Executive and Legislative branches of the Government must work together.

This is a serious course upon which we embark.

Excerpt from "The Truman Doctrine" by Harry S. Truman

21. What is the primary purpose of this speech?
    a. To persuade
    b. To describe
    c. To entertain
    d. To analyze

22. After hearing this speech, one could infer Truman believes that the United States is _____ and Greece and Turkey are _____.
    a. A developed nation; third-world countries
    b. Part of the United Nations; not part of the United Nations
    c. A free democracy; under dictatorships
    d. An altruistic hero; potential victims of oppression

23. President Truman describes two ways of life. Which statement corresponds to his description?
    a. The first way of life is distinguished by free institutions; the second is marked by freedom of religion.
    b. The first way of life relies on terror and oppression; the second controls the press and radio.
    c. The first way of life guarantees individual liberties; the second suppresses personal freedoms.
    d. The first way of life suppresses individual liberties; the second guarantees personal freedoms.

24. What does the underlined word mean in the following sentence?

    "But we cannot allow changes in the status quo in violation of the Charter of the United Nations by such methods as coercion, or by such <u>subterfuges</u> as political infiltration."

    a. Helpful pieces of intel
    b. Tricky deceptions
    c. Time-consuming activities
    d. Underground routes

25. Based on the speech, what would be an immediate effect of Greece falling under the control of an armed minority?
    a. Turkey and the entire Middle East might experience unprecedented turmoil.
    b. The Charter of the United Nations would prevent any negative repercussions.
    c. Greece would receive hundreds of millions of dollars from the United States.
    d. Greece might become absorbed into another country and disappear off the map completely.

26. Truman states, "Should we fail to aid Greece and Turkey in this fateful hour, the effect will be far reaching to the West as well as to the East." What emotion might members of Congress feel at this moment?
    a. Disgust
    b. Astonishment
    c. Concern
    d. Eagerness

27. What is NOT one of Truman's requests to Congress?
    a. Send four million dollars in relief funds to Greece and Turkey.
    b. Overhaul the governments of Greece and Turkey.
    c. Deploy civil personnel to Greece and Turkey.
    d. Authorize the fastest and most effective use of supplies and equipment.

28. If Congress does not agree to Truman's plea, what is likely to happen?
    a. Another country will step up and offer similar relief aid.
    b. Greece and Turkey will rely on each other to avoid suppression.
    c. Most Eastern countries will experience the collapse of institutions and freedom.
    d. The United Nations will force the U.S. Congress to accept Truman's terms.

29. Which quote indicates that time is of the essence?
    a. "The choice is too often not a free one."
    b. "The world is not static, and the status quo is not sacred."
    c. "This is a serious course upon which we embark."
    d. "We must take immediate and resolute action."

30. How is this speech organized?
    a. Problem and solution
    b. Chronological order
    c. Compare and contrast
    d. Categorical

Science

Questions 31–42 are based on the following two passages:

Passage 1

"Insects are the flower's auxiliaries. Flies, wasps, honeybees, bumblebees, beetles, butterflies, all vie with one another in rendering aid by carrying the pollen of the stamens to the stigmas. They dive into the flower, enticed by a honeyed drop expressly prepared at the bottom of the corolla. In their efforts to obtain it they shake the stamens and daub themselves with pollen, which they carry from one flower to another. Who has not seen bumblebees coming out of the bosom of the flowers all covered with pollen? Their hairy stomachs, powdered with pollen, have only to touch a stigma in passing to communicate life to it. When in the spring you see on a blooming pear-tree, a whole swarm of flies, bees, and butterflies, hurrying, humming, and fluttering, it is a triple feast, my friends: a feast for the insect that pilfers in the depth of the flowers; a feast for the tree whose ovaries are quickened by all these merry little people; and a feast for man, to whom abundant harvest is promised. The insect is the best distributor of pollen. All the flowers it visits receive their share of quickening dust."

[...]

"To attract the insect that it needs, every flower has at the bottom of its corolla a drop of sweet liquor called nectar. From this liquor bees make their honey. To draw it from corollas shaped like a deep funnel, butterflies have a long trumpet, curled in a spiral when at rest, but which they unroll and plunge into the flower like a bore when they wish to obtain the delicious drink. The insect does not see this drop of nectar; however, it knows that it is there and finds it without hesitation. But in some flowers a grave difficulty presents itself: these flowers are closed tight everywhere. How is the treasure to be got at, how find the entrance that leads to the nectar? Well, these closed flowers have a signboard, a mark that says clearly: Enter here."

"You won't make us believe that!" cried Claire.

"I am not going to make you believe anything, my dear child; I am going to show you. Look at this snapdragon blossom. It is shut tight, its two closed lips leave no passage between. Its color is a uniform purplish red; but there, just in the middle of the lower lip, is a large spot of bright yellow. This spot, so appropriate for catching the eye, is the mark, the signboard I told you of. By its brightness it says: Here is the keyhole.

"Press your little finger on the spot. You see. The flower yawns immediately, the secret lock works. And you think the bumblebee does not know these things? Watch it in the garden and you will see how it can read the signs of the flowers. When it visits a snapdragon, it always alights on the yellow spot and nowhere else. The door opens, it enters. It twists and turns in the

17

corolla and covers itself with pollen, with which it daubs the stigma. Having drunk the drop, it goes off to other flowers, forcing the opening of which it knows the secret thoroughly.

"All closed flowers have, like the snapdragon, a conspicuous point, a spot of bright color, a sign that shows the insect the entrance to the corolla and says to it: Here it is. Finally, insects whose trade it is to visit flowers and make the pollen fall from the stamens on to the stigma, have a wonderful knowledge of the significance of this spot. It is on it they use their strength to make the flower open.

"Let us recapitulate. Insects are necessary to flowers to bring pollen to the stigmas. A drop of nectar, distilled on purpose for this, attracts them to the bottom of the corolla; a bright spot shows them the road to follow. Either I am a triple idiot or we have here an admirable chain of facts. Later, my children, you will find only too many people saying: This world is the product of chance, no intelligence rules it, no Providence guides it. To those people, my friends, show the snapdragon's yellow spot. If, less clear-sighted than the burly bumblebee, they do not understand it, pity them: they have diseased brains."

Excerpt from The Storybook of Science by Jean-Henri Fabre

Passage 2

A very complete knowledge of the pollen-gathering behavior of the worker honeybee may be obtained by a study of the actions of bees which are working upon a plant which yields pollen in abundance. Sweet corn is an ideal plant for this purpose, and it will be used as a basis for the description which follows.

The movements of the legs and of the mouthparts are so rapid and so many members are in action at once that it is impossible for the eye to follow all at the same time. However, long-continued observation, assisted by the study of instantaneous photographs, gives confidence that the statements recorded are accurate, although some movements may have escaped notice.

To obtain pollen from corn the bee must find a tassel in the right stage of ripeness, with flowers open and stamens hanging from them. The bee alights upon a spike and crawls along it, clinging to the pendent anthers. It crawls over the anthers, going from one flower to another along the spike, being all the while busily engaged in the task of obtaining pollen. This reaches its body in several ways.

As the bee moves over the anthers it uses its mandibles and tongue, biting the anthers and licking them and securing a considerable amount of pollen upon these parts. This pollen becomes moist and sticky, since it is mingled with fluid from the mouth. A considerable amount of pollen is dislodged from the anthers as the bee moves over them. All of the legs receive a supply of this free pollen and much adheres to the hairs which cover the body, more particularly to those upon the ventral surface. This free pollen is dry and powdery and is very different in appearance from the moist pollen masses with which the bee returns to the hive. Before the return journey this pollen must be transferred to the baskets and securely packed in them.

After the bee has traversed a few flowers along the spike and has become well supplied with free pollen it begins to collect it from its body, head, and forward appendages and to transfer it to the posterior pair of legs. This may be accomplished while the bee is resting upon the flower or while it is hovering in the air before seeking additional pollen. It is probably more thoroughly and rapidly accomplished while the bee is in the air, since all of the legs are then free to function in the gathering process.

If the collecting bee is seized with forceps and examined after it has crawled over the stamens of a few flowers of the corn, its legs and the ventral surface of its body are found to be thickly powdered over with pollen. If the bee hovers in the air for a few moments and is then examined very little pollen is found upon the body or upon the legs, except the masses within the pollen baskets. While in the air it has accomplished the work of collecting some of the scattered grains and of storing them in the baskets, while others have been brushed from the body.

In attempting to describe the movements by which this result is accomplished it will be best first to sketch briefly the roles of the three pairs of legs. They are as follows:

(a) The first pair of legs remove scattered pollen from the head and the region of the neck, and the pollen that has been moistened by fluid substances from the mouth.

(b) The second pair of legs remove scattered pollen from the thorax, more particularly from the ventral region, and they receive the pollen that has been collected by the first pair of legs.

(c) The third pair of legs collect a little of the scattered pollen from the abdomen and they receive pollen that has been collected by the second pair. Nearly all of this pollen is collected by the pollen combs of the hind legs, and is transferred from the combs to the pollen baskets or corbiculae in a manner to be described later.

It will thus be seen that the manipulation of pollen is a successive process, and that most of the pollen at least passes backward from the point where it happens to touch the bee until it finally reaches the corbiculae or is accidentally dislodged and falls from the rapidly moving limbs.

Excerpt from The Behavior of the Honey Bee in Pollen Collection by Dana Brackenridge Casteel

31. What do these two articles have in common?
    a. Both articles describe the roles of the bee's three different sets of legs.
    b. Both articles articulate the important role insects play in pollination.
    c. Both articles present their findings in a formal and scientific manner.
    d. Both articles discuss the methods that flowers use to attract pollinators.

32. The diction Jean-Henri Fabre uses in the first passage is:
    a. Pedantic
    b. Emotional
    c. Whimsical
    d. Ambiguous

33. What does the underlined word mean in the following sentence?

"After the bee has traversed a few flowers along the spike and has become well supplied with free pollen it begins to collect it from its body, head, and forward appendages and to transfer it to the <u>posterior</u> pair of legs."

    a. Strongest
    b. Anterior
    c. Ambulatory
    d. Rear

34. Which type of evidence is used by the author of the second passage, Dana Brackenridge?
    a. Anecdotal
    b. Hypothetical
    c. Experimental
    d. Observational

35. In the first passage, what conclusion can be drawn about this character's feeling toward nature from the following sentence?

"To those people, my friends, show the snapdragon's yellow spot."

    a. The character believes nature is designed with purpose and intent.
    b. The character believes nature is vibrant yet random.
    c. The character believes nature's most beautiful plant is the snapdragon.
    d. The character believes nature is difficult to predict and understand.

36. Which entity does NOT benefit from the "triple feast?"
    a. Trees
    b. Insects
    c. Pollen
    d. Humans

37. Pollen manipulation by the honeybee is a successive process that:
    a. Starts in the pollen basket and moves toward the legs
    b. Starts on the body, moves to the legs, and eventually ends at the corbiculae
    c. Starts on the thorax and moves toward the tongue and mandibles
    d. Starts on the antennae, moves to the wings, and eventually ends at the pollen basket

38. Which two terms are synonyms?
    a. Nectar; honey
    b. Stamen; stigma
    c. Pollen baskets; corbiculae
    d. Thorax; abdomen

39. Which kind of logical fallacy is presented in the following sentence?

"Either I am a triple idiot or we have here an admirable chain of facts."

   a. Slippery slope
   b. Ad hominem
   c. Red herring
   d. False dichotomy

40. In paragraph five of the first passage, what speaker imply when he says the "flower yawns"?
   a. The petals of the flower are opening up slowly.
   b. The flower is worn out from pollination and needs rest.
   c. The flower is expanding its surface area to absorb more carbon dioxide.
   d. The flower is having a difficult time regulating night and day.

41. Why does the study in the second passage use sweet corn?
   a. Honeybees are attracted to the natural sweetness of the plant.
   b. The region has an abundance of sweet corn available for study.
   c. Sweet corn does not attract any other species of pollinator, making it easier to focus on bees.
   d. Sweet corn yields a generous amount of pollen.

42. What conclusion can be drawn after reading both of these passages?
   a. To date, science has had very to say little about bees, flowers, and pollination.
   b. Honeybees and flowers have a mutually beneficial relationship.
   c. Bees are the most advanced pollinator.
   d. The process of pollination is more accidental than intentional.

Science

Questions 43–52 are based upon the following passage:

How does soap make your hands clean?

Why will gasoline take a grease spot out of your clothes?

If we were to go back to our convenient imaginary switchboard to turn off another law, we should find near the heat switches, and not far from the chemistry ones, a switch labeled Solution. Suppose we turned it off:

The fishes in the sea are among the first creatures to be surprised by our action. For instantly all the salt in the ocean drops to the bottom like so much sand, and most saltwater fishes soon perish in the freshwater.

[...]

Probably we had better let the Solution switch alone, after all. Instead, here are a couple of experiments that will help to make clear what happens when anything dissolves to make a solution.

Experiment 80. Fill a test tube one fourth full of cold water. Slowly stir in salt until no more will dissolve. Add half a teaspoonful more of salt than will dissolve. Dry the outside of the test tube

21

and heat the salty water over the Bunsen burner. Will hot water dissolve things more readily or less readily than cold? Why do you wash dishes in hot water?

Experiment 81. Fill a test tube one fourth full of any kind of oil, and one fourth full of water. Hold your thumb over the top of the test tube and shake it hard for a minute or two. Now look at it. Pour it out, and shake some prepared cleanser into the test tube, adding a little more water. Shake the test tube thoroughly and rinse. Put it away clean.

When you shake the oil with the water, the oil breaks up into tiny droplets. These droplets are so small that they reflect the light that strikes them and so look white, or pale yellow. This milky mixture is called an emulsion. Milk is an emulsion; there are tiny droplets of butter fat and other substances scattered all through the milk. The butter fat is not dissolved in the rest of the milk, and the oil is not dissolved in the water. But the droplets may be so small that an emulsion acts almost exactly like a solution.

But when you shake or stir salt or sugar in water, the particles divide up into smaller and smaller pieces, until probably each piece is just a single molecule of the salt or sugar. And these molecules get into the spaces between the water molecules and bounce around among them. They therefore act like the water and let the light through. This is a solution. The salt or sugar is dissolved in the water. Any liquid mixture which remains clear is a solution, no matter what the color. Most red ink, most blueing, clear coffee, tea, and ocean water are solutions. If a liquid is clear, no matter what the color, you can be sure that whatever things may be in it are dissolved.

Experiment 82. Pour alcohol into a test tube (square-bottomed test tubes are best for this experiment), standing the tube up beside a ruler. When the alcohol is just one inch high in the tube, stop pouring. Put exactly the same amount of water in another test tube of the same size. When you pour them together, how many inches high do you think the mixture will be? Pour the water into the alcohol, shake the mixture a little, and measure to see how high it comes in the test tube. Did you notice the warmth when you shook the tube?

If you use denatured alcohol, you are likely to have an emulsion as a result of the mixing. The alcohol part of the denatured alcohol dissolves in the water well enough, but the denaturing substance in the alcohol will not dissolve in water; so it forms tiny droplets that make the mixture of alcohol and water cloudy.

The purpose of this experiment is to show that the molecules of water get into the spaces between the molecules of alcohol. It is as if you were to add a pail of pebbles to a pail of apples. The pebbles would fill in between the apples, and the mixture would not nearly fill two pails.

Excerpt from "Common Science" by Carleton W. Washburne

43. Which title best suits this chapter?
    a. Cleaning with Science
    b. Experiments You Can Do at Home
    c. Solutions and Emulsions
    d. Solution Switchboard

44. What effect does the author achieve by presenting the scenario of the imaginary?
    a. Humor through a hypothetical situation
    b. Shock through a surprise twist
    c. Deception through misleading information
    d. Curiosity through growing suspense

45. Why does the author ask rhetorical questions?
    a. To reveal their opinion on the topic
    b. To encourage the reader to draw their own conclusions
    c. To ensure the reader is following along
    d. To convince the reader to conduct the experiments at home

46. How are emulsions different from solutions?
    a. Emulsions are clear; solutions are cloudy.
    b. Emulsions are non-potable; solutions are potable.
    c. Emulsions have a low temperature; solutions have a high temperature.
    d. Emulsions have tiny, suspended droplets; solutions are completely dissolved.

47. What is the most likely result of Experiment 80?
    a. The heat was unable to dissolve the additional salt.
    b. Once heated, the additional salt dissolved completely.
    c. The additional salt prevented the water from heating up more than room temperature.
    d. The heated salt and water solution could be used to wash dishes.

48. When might an emulsion pass as a solution?
    a. When the scattered droplets cannot be detected because they are so small.
    b. When the substance is heated to its boiling point.
    c. When the emulsion has been stirred or shaken thoroughly.
    d. Emulsions cannot pass as solutions.

49. Which analogy makes the correct comparison?
    a. Pebbles are to apples as alcohol is to water.
    b. Water is to pebbles as alcohol is to apples.
    c. Apples are to water as pebbles are to alcohol.
    d. Alcohol and water are to emulsion as apples and pebbles are to solution.

50. What can be inferred about denatured alcohol?
    a. It can be dangerous to work with in a laboratory.
    b. It has the same basic properties as regular alcohol.
    c. It needs to be heated in order to completely dissolve.
    d. There are better alternatives to form true solutions.

51. Using deductive reasoning, how high will the mixture measure at the end of Experiment 82?
    a. Exactly two inches
    b. More than two inches
    c. Less than two inches
    d. There is no way to tell unless you conduct the experiment.

52. Which idiom could be explained using evidence from this passage?
    a. That is like comparing apples to oranges.
    b. She is going to be in hot water.
    c. They are like oil and water.
    d. No use crying over spilled milk.

Writing and Language Test

Questions 1–9 are based on the following passage:

While all dogs (1) <u>descend through gray wolves</u>, it's easy to notice that dog breeds come in a variety of shapes and sizes. With such a (2) <u>drastic range of traits, appearances and body types</u> dogs are one of the most variable and adaptable species on the planet. (3) <u>But why so many differences.</u> The answer is that humans have actually played a major role in altering the biology of dogs. (4) <u>This was done through a process called selective breeding.</u>

(5) <u>Selective breeding which is also called artificial selection is the process</u> in which animals with desired traits are bred in order to produce offspring that share the same traits. In natural evolution, (6) <u>animals must adapt to their environments increase their chance of survival.</u> Over time, certain traits develop in animals that enable them to thrive in these environments. Those animals with more of these traits, or better versions of these traits, gain an (7) <u>advantage over others of their species.</u> Therefore, the animal's chances to mate are increased and these useful (8) <u>genes are passed into their offspring.</u> With dog breeding, humans select traits that are desired and encourage more of these desired traits in other dogs by breeding dogs that already have them.

The reason for different breeds of dogs is that there were specific needs that humans wanted to fill with their animals. For example, scent hounds are known for their extraordinary ability to track game through scent. These breeds are also known for their endurance in seeking deer and other prey. Therefore, early hunters took dogs that displayed these abilities and bred them to encourage these traits. Later generations took on characteristics that aided these desired traits. (9) <u>For example, Bloodhounds</u> have broad snouts and droopy ears that fall to the ground when they smell. These physical qualities not only define the look of the bloodhound, but also contribute to their amazing tracking ability. The broad snout is able to define and hold onto scents longer than many other breeds. The long, floppy ears serve to collect and hold the scents the earth holds so that the smells are clearer and able to be distinguished.

1. Which of the following would be the best choice for this sentence (reproduced below)?

While all dogs (1) <u>descend through gray wolves</u>, it's easy to notice that dog breeds come in a variety of shapes and sizes.

a. NO CHANGE
b. descend by gray wolves
c. descend from gray wolves
d. descended through gray wolves

2. Which of the following would be the best choice for this sentence (reproduced below)?

With such a (2) drastic range of traits, appearances and body types, dogs are one of the most variable and adaptable species on the planet.

a. NO CHANGE
b. drastic range of traits, appearances, and body types,
c. drastic range of traits and appearances and body types,
d. drastic range of traits, appearances, as well as body types,

3. Which of the following would be the best choice for this sentence (reproduced below)?

(3) But why so many differences.

a. NO CHANGE
b. But are there so many differences?
c. But why so many differences are there.
d. But why are there so many differences?

4. Which of the following would be the best choice for this sentence (reproduced below)?

(4) This was done through a process called selective breeding.

a. NO CHANGE
b. This was done, through a process called selective breeding.
c. This was done, through a process, called selective breeding.
d. This was done through selective breeding, a process.

5. Which of the following would be the best choice for this sentence (reproduced below)?

(5) Selective breeding which is also called artificial selection is the process in which animals with desired traits are bred in order to produce offspring that share the same traits.

a. NO CHANGE
b. Selective breeding, which is also called artificial selection is the process
c. Selective breeding which is also called, artificial selection, is the process
d. Selective breeding, which is also called artificial selection, is the process

6. Which of the following would be the best choice for this sentence (reproduced below)?

In natural evolution, (6) <u>animals must adapt to their environments increase their chance of survival.</u>

a. NO CHANGE
b. animals must adapt to their environments to increase their chance of survival.
c. animals must adapt to their environments, increase their chance of survival.
d. animals must adapt to their environments, increasing their chance of survival.

7. Which of the following would be the best choice for this sentence (reproduced below)?

Those animals with more of these traits, or better versions of these traits, gain an (7) <u>advantage over others of their species.</u>

a. NO CHANGE
b. advantage over others, of their species.
c. advantages over others of their species.
d. advantage over others.

8. Which of the following would be the best choice for this sentence (reproduced below)?

Therefore, the animal's chances to mate are increased and these useful (8) <u>genes are passed into their offspring.</u>

a. NO CHANGE
b. genes are passed onto their offspring.
c. genes are passed on to their offspring.
d. genes are passed within their offspring.

9. Which of the following would be the best choice for this sentence (reproduced below)?

(9) <u>For example, Bloodhounds</u> have broad snouts and droopy ears that fall to the ground when they smell.

a. NO CHANGE
b. For example, Bloodhounds,
c. For example Bloodhounds
d. For example, bloodhounds

Questions 10–18 are based on the following passage:

I'm not alone when I say that it's hard to pay attention sometimes. I can't count how many times I've sat in a classroom, lecture, speech, or workshop and (10) <u>been bored to tears or rather sleep.</u> (11) <u>Usually I turn to doodling in order to keep awake.</u> This never really helps; I'm not much of an artist. Therefore, after giving up on drawing a masterpiece, I would just concentrate on keeping my eyes open and trying to be attentive. This didn't always work because I wasn't engaged in what was going on.

(12) <u>Sometimes in particularly dull seminars,</u> I'd imagine comical things going on in the room or with the people trapped in the room with me. Why? (13) <u>Because I wasn't invested in what was</u>

going on I wasn't motivated to listen. I'm not going to write about how I conquered the difficult task of actually paying attention in a difficult or unappealing class—it can be done, sure. I have sat through the very epitome of boredom (in my view at least) several times and come away learning something. (14) Everyone probably has had to at one time do this. What I want to talk about is that profound moment when curiosity is sparked (15) in another person drawing them to pay attention to what is before them and expand their knowledge.

What really makes people pay attention? (16) Easy it's interest. This doesn't necessarily mean (17) embellishing subject matter drawing people's attention. This won't always work. However, an individual can present material in a way that is clear to understand and actually engages the audience. Asking questions to the audience or class will make them a part of the topic at hand. Discussions that make people think about the content and (18) how it applies to there lives world and future are key. If math is being discussed, an instructor can explain the purpose behind the equations or perhaps use real-world applications to show how relevant the topic is. When discussing history, a lecturer can prompt students to imagine themselves in the place of key figures and ask how they might respond. The bottom line is to explore the ideas rather than just lecture. Give people the chance to explore material from multiple angles, and they'll be hungry to keep paying attention for more information.

10. Which of the following would be the best choice for this sentence (reproduced below)?

I can't count how many times I've sat in a classroom, lecture, speech, or workshop and (10) been bored to tears or rather sleep.

a. NO CHANGE
b. been bored to, tears, or rather sleep.
c. been bored, to tears or rather sleep.
d. been bored to tears or, rather, sleep.

11. Which of the following would be the best choice for this sentence (reproduced below)?

(11) Usually I turn to doodling in order to keep awake.

a. NO CHANGE
b. Usually, I turn to doodling in order to keep awake.
c. Usually I turn to doodling, in order, to keep awake.
d. Usually I turned to doodling in order to keep awake.

12. Which of the following would be the best choice for this sentence (reproduced below)?

(12) Sometimes in particularly dull seminars, I'd imagine comical things going on in the room or with the people trapped in the room with me.

a. NO CHANGE
b. Sometimes, in particularly, dull seminars,
c. Sometimes in particularly dull seminars
d. Sometimes in particularly, dull seminars,

13. Which of the following would be the best choice for this sentence (reproduced below)?

(13) <u>Because I wasn't invested in what was going on I wasn't motivated to listen.</u>

a. NO CHANGE
b. Because I wasn't invested, in what was going on, I wasn't motivated to listen.
c. Because I wasn't invested in what was going on. I wasn't motivated to listen.
d. I wasn't motivated to listen because I wasn't invested in what was going on.

14. Which of the following would be the best choice for this sentence (reproduced below)?

(14) <u>Everyone probably has had to at one time do this.</u>

a. NO CHANGE
b. Everyone probably has had to, at one time. Do this.
c. Everyone's probably had to do this at one time.
d. At one time everyone probably has had to do this.

15. Which of the following would be the best choice for this sentence (reproduced below)?

What I want to talk about is that profound moment when curiosity is sparked (15) <u>in another person drawing them to pay attention to what is before them</u> and expand their knowledge.

a. NO CHANGE
b. in another person, drawing them to pay attention
c. in another person; drawing them to pay attention to what is before them.
d. in another person, drawing them to pay attention to what is before them.

16. Which of the following would be the best choice for this sentence (reproduced below)?

(16) <u>Easy it's interest.</u>

a. NO CHANGE
b. Easy it is interest.
c. Easy. It's interest.
d. Easy—it's interest.

17. Which of the following would be the best choice for this sentence (reproduced below)?

This doesn't necessarily mean (17) <u>embellishing subject matter drawing people's attention.</u>

a. NO CHANGE
b. embellishing subject matter which draws people's attention.
c. embellishing subject matter to draw people's attention.
d. embellishing subject matter for the purpose of drawing people's attention.

18. Which of the following would be the best choice for this sentence (reproduced below)?

Discussions that make people think about the content and (18) how it applies to there lives world and future are key.

a. NO CHANGE
b. how it applies to their lives, world, and future are key.
c. how it applied to there lives world and future are key.
d. how it applies to their lives, world and future are key.

Questions 19–27 are based on the following passage:

Since the first discovery of dinosaur bones, (19) scientists has made strides in technological development and methodologies used to investigate these extinct animals. We know more about dinosaurs than ever before and are still learning fascinating new things about how they looked and lived. However, one has to ask, (20) how if earlier perceptions of dinosaurs continue to influence people's understanding of these creatures? Can these perceptions inhibit progress towards further understanding of dinosaurs?

(21) The biggest problem with studying dinosaurs is simply that there are no living dinosaurs to observe. All discoveries associated with these animals are based on physical remains. To gauge behavioral characteristics, scientists cross-examine these (22) finds with living animals that seem similar in order to gain understanding. While this method is effective, these are still deductions. Some ideas about dinosaurs can't be tested and confirmed simply because humans can't replicate a living dinosaur. For example, a Spinosaurus has a large sail, or a finlike structure that grows from its back. Paleontologists know this sail exists and have ideas for the function of (23) the sail however they are uncertain of which idea is the true function. Some scientists believe (24) the sail serves to regulate the Spinosaurus' body temperature and yet others believe its used to attract mates. Still, other scientists think the sail is used to intimidate other predatory dinosaurs for self-defense. These are all viable explanations, but they are also influenced by what scientists know about modern animals. (25) Yet, it's quite possible that the sail could hold a completely unique function.

While it's (26) plausible, even likely that dinosaurs share many traits with modern animals, there is the danger of overattributing these qualities to a unique, extinct species. For much of the early nineteenth century, when people first started studying dinosaur bones, the assumption was that they were simply giant lizards. (27) For the longest time this image was the prevailing view on dinosaurs, until evidence indicated that they were more likely warm blooded. Scientists have also discovered that many dinosaurs had feathers and actually share many traits with modern birds.

19. Which of the following would be the best choice for this sentence (reproduced below)?

Since the first discovery of dinosaur bones, (19) <u>scientists has made strides in technological development and methodologies used to investigate</u> these extinct animals.

a. NO CHANGE
b. scientists has made strides in technological development, and methodologies, used to investigate
c. scientists have made strides in technological development and methodologies used to investigate
d. scientists, have made strides in technological development and methodologies used, to investigate

20. Which of the following would be the best choice for this sentence (reproduced below)?

However, one has to ask, (20) <u>how if earlier perceptions of dinosaurs</u> continue to influence people's understanding of these creatures?

a. NO CHANGE
b. how perceptions of dinosaurs
c. how, if, earlier perceptions of dinosaurs
d. whether earlier perceptions of dinosaurs

21. Which of the following would be the best choice for this sentence (reproduced below)?

(21) <u>The biggest problem with studying dinosaurs is simply that there are no living dinosaurs to observe.</u>

a. NO CHANGE
b. The biggest problem with studying dinosaurs is simple, that there are no living dinosaurs to observe.
c. The biggest problem with studying dinosaurs is simple. There are no living dinosaurs to observe.
d. The biggest problem with studying dinosaurs, is simply that there are no living dinosaurs to observe.

22. Which of the following would be the best choice for this sentence (reproduced below)?

To gauge behavioral characteristics, scientists cross-examine these (22) <u>finds with living animals that seem similar in order to gain understanding.</u>

a. NO CHANGE
b. finds with living animals to explore potential similarities.
c. finds with living animals to gain understanding of similarities.
d. finds with living animals that seem similar, in order, to gain understanding.

23. Which of the following would be the best choice for this sentence (reproduced below)?

Paleontologists know this sail exists and have ideas for the function of (23) <u>the sail however they are uncertain of which idea is the true function.</u>

a. NO CHANGE
b. the sail however, they are uncertain of which idea is the true function.
c. the sail however they are, uncertain, of which idea is the true function.
d. the sail; however, they are uncertain of which idea is the true function.

24. Which of the following would be the best choice for this sentence (reproduced below)?

Some scientists believe (24) <u>the sail serves to regulate the Spinosaurus' body temperature and yet others believe its used to attract mates.</u>

a. NO CHANGE
b. the sail serves to regulate the Spinosaurus' body temperature, yet others believe it's used to attract mates.
c. the sail serves to regulate the Spinosaurus' body temperature and yet others believe it's used to attract mates.
d. the sail serves to regulate the Spinosaurus' body temperature however others believe it's used to attract mates.

25. Which of the following would be the best choice for this sentence (reproduced below)?

(25) <u>Yet, it's quite possible</u> that the sail could hold a completely unique function.

a. NO CHANGE
b. Yet, it's quite possible,
c. It's quite possible,
d. Its quite possible

26. Which of the following would be the best choice for this sentence (reproduced below)?

While it's (26) <u>plausible, even likely that dinosaurs share many</u> traits with modern animals, there is the danger of over attributing these qualities to a unique, extinct species.

a. NO CHANGE
b. plausible, even likely that, dinosaurs share many
c. plausible, even likely, that dinosaurs share many
d. plausible even likely that dinosaurs share many

27. Which of the following would be the best choice for this sentence (reproduced below)?

(27) <u>For the longest time this image was the prevailing view on dinosaurs</u>, until evidence indicated that they were more likely warm blooded.

a. NO CHANGE
b. For the longest time this was the prevailing view on dinosaurs
c. For the longest time, this image, was the prevailing view on dinosaurs
d. For the longest time this was the prevailing image of dinosaurs

Questions 28–36 are based on the following passage written about Perelandra *by C.S. Lewis:*

Everyone has heard the (28) <u>idea of the end justifying the means; that would be Weston's philosophy.</u> Weston is willing to cross any line, commit any act no matter how heinous, to achieve success in his goal. (29) <u>Ransom is reviled by this fact, seeing total evil in Weston's plan.</u> To do an evil act in order (30) <u>to gain a result that's supposedly good would ultimately warp the final act.</u> (31) <u>This opposing viewpoints immediately distinguishes Ransom as the hero.</u> In the conflict with Un-man, Ransom remains true to his moral principles, someone who refuses to be compromised by power. Instead, Ransom makes it clear that by allowing such processes as murder and lying to dictate how one attains a positive outcome, (32) <u>the righteous goal becomes corrupted.</u> The good end would not be truly good, but a twisted end that conceals corrupt deeds.

(33) <u>This idea of allowing necessary evils to happen, is very tempting, it is what Weston fell prey to.</u> (34) <u>The temptation of the evil spirit Un-man ultimately takes over Weston and he is possessed.</u> However, Ransom does not give into temptation. He remains faithful to the truth of what is right and incorrect. This leads him to directly face Un-man for the fate of Perelandra and its inhabitants.

Just as Weston was corrupted by the Un-man, (35) <u>Un-man after this seeks to tempt the Queen of Perelandra</u> to darkness. Ransom must literally (36) <u>show her the right path, to accomplish this, he does this based on the same principle as the "means to an end" argument</u>—that good follows good, and evil follows evil. Later in the plot, Weston/Un-man seeks to use deceptive reasoning to turn the queen to sin, pushing the queen to essentially ignore Melildil's rule to satisfy her own curiosity. In this sense, Un-man takes on the role of a false prophet, a tempter. Ransom must shed light on the truth, but this is difficult; his adversary is very clever and uses brilliant language. Ransom's lack of refinement heightens the weight of Un-man's corrupted logic, and so the Queen herself is intrigued by his logic.

28. Which of the following would be the best choice for this sentence (reproduced below)?

Everyone has heard the (28) <u>idea of the end justifying the means; that would be Weston's philosophy.</u>

a. NO CHANGE
b. idea of the end justifying the means; this is Weston's philosophy.
c. idea of the end justifying the means, this is the philosophy of Weston
d. idea of the end justifying the means. That would be Weston's philosophy.

29. Which of the following would be the best choice for this sentence (reproduced below)?

(29) Ransom is reviled by this fact, seeing total evil in Weston's plan.

a. NO CHANGE
b. Ransom is reviled by this fact; seeing total evil in Weston's plan.
c. Ransom, is reviled by this fact, seeing total evil in Weston's plan.
d. Ransom reviled by this, sees total evil in Weston's plan.

30. Which of the following would be the best choice for this sentence (reproduced below)?

To do an evil act in order (30) to gain a result that's supposedly good would ultimately warp the final act.

a. NO CHANGE
b. for an outcome that's for a greater good would ultimately warp the final act.
c. to gain a final act would warp its goodness.
d. to achieve a positive outcome would ultimately warp the goodness of the final act.

31. Which of the following would be the best choice for this sentence (reproduced below)?

(31) This opposing viewpoints immediately distinguishes Ransom as the hero.

a. NO CHANGE
b. This opposing viewpoints immediately distinguishes Ransom, as the hero.
c. This opposing viewpoint immediately distinguishes Ransom as the hero.
d. Those opposing viewpoints immediately distinguishes Ransom as the hero.

32. Which of the following would be the best choice for this sentence (reproduced below)?

Instead, Ransom makes it clear that by allowing such processes as murder and lying to dictate how one attains a positive outcome, (32) the righteous goal becomes corrupted.

a. NO CHANGE
b. the goal becomes corrupted and no longer righteous.
c. the righteous goal becomes, corrupted.
d. the goal becomes corrupted, when once it was righteous.

33. Which of the following would be the best choice for this sentence (reproduced below)?

(33) This idea of allowing necessary evils to happen, is very tempting, it is what Weston fell prey to.

a. NO CHANGE
b. This idea of allowing necessary evils to happen, is very tempting. This is what Weston fell prey to.
c. This idea, allowing necessary evils to happen, is very tempting, it is what Weston fell prey to.
d. This tempting idea of allowing necessary evils to happen is what Weston fell prey to.

34. Which of the following would be the best choice for this sentence (reproduced below)?

(34) <u>The temptation of the evil spirit Un-man ultimately takes over Weston and he is possessed.</u>

a. NO CHANGE
b. The temptation of the evil spirit Un-man ultimately takes over and possesses Weston.
c. Weston is possessed as a result of the temptation of the evil spirit Un-man ultimately, who takes over.
d. The temptation of the evil spirit Un-man takes over Weston and he is possessed ultimately.

35. Which of the following would be the best choice for this sentence (reproduced below)?

Just as Weston was corrupted by the Un-man, (35) <u>Un-man after this seeks to tempt the Queen of Perelandra</u> to darkness.

a. NO CHANGE
b. Un-man, after this, would tempt the Queen of Perelandra
c. Un-man, after this, seeks to tempt the Queen of Perelandra
d. Un-man then seeks to tempt the Queen of Perelandra

36. Which of the following would be the best choice for this sentence (reproduced below)?

Ransom must literally (36) <u>show her the right path, to accomplish this, he does this based on the same principle as the "means to an end" argument</u>—that good follows good, and evil follows evil.

a. NO CHANGE
b. show her the right path. To accomplish this, he uses the same principle as the "means to an end" argument
c. show her the right path; to accomplish this he uses the same principle as the "means to an end" argument
d. show her the right path, to accomplish this, the same principle as the "means to an end" argument is applied

Questions 37–45 are based on the following passage:

(37) <u>What's clear about the news today is that the broader the media</u> the more ways there are to tell a story. Even if different news groups cover the same story, individual newsrooms can interpret or depict the story differently than other counterparts. Stories can also change depending on the type of (38) <u>media in question incorporating different styles and unique</u> ways to approach the news. (39) <u>It is because of these respective media types that ethical and news-related subject matter can sometimes seem different or altered.</u> But how does this affect the narrative of the new story?

I began by investigating a written newspaper article from the Baltimore Sun. Instantly striking are the bolded headlines. (40) <u>These are clearly meant for direct the viewer</u> to the most exciting and important stories the paper has to offer. What was particularly noteworthy about this edition was that the first page dealt with two major ethical issues. (41) <u>On a national level there was a story</u> on the evolving Petraeus scandal involving his supposed affair. The other article was focused locally in Baltimore, a piece questioning the city's Ethics Board and their current director. Just as a television newscaster communicates the story through camera and dialogue, the printed article applies intentional and targeted written narrative style. More so than any of the mediums, a news article seems to be focused specifically on a given story without need to jump to another. Finer details are usually expanded on (42) <u>in written articles, usually people who</u> read newspapers or go online for web articles want more than a quick blurb. The diction of the story is also more precise and can be either straightforward or suggestive (43) <u>depending in earnest on the goal of the writer.</u> However, there's still plenty of room for opinions to be inserted into the text.

Usually, all news (44) <u>outlets have some sort of bias, it's just a question of how much</u> bias clouds the reporting. As long as this bias doesn't withhold information from the reader, it can be considered credible. (45) <u>However an over use of bias</u>, opinion, and suggestive language can rob readers of the chance to interpret the news events for themselves.

37. Which of the following would be the best choice for this sentence (reproduced below)?

(37) <u>What's clear about the news today is that the broader the media</u> the more ways there are to tell a story.

a. NO CHANGE
b. What's clear, about the news today, is that the broader the media
c. What's clear about today's news is that the broader the media
d. The news today is broader than earlier media

38. Which of the following would be the best choice for this sentence (reproduced below)?

Stories can also change depending on the type of (38) <u>media in question incorporating different styles and unique</u> ways to approach the news.

a. NO CHANGE
b. media in question; each incorporates unique styles and unique
c. media in question. To incorporate different styles and unique
d. media in question, incorporating different styles and unique

39. Which of the following would be the best choice for this sentence (reproduced below)?

(39) <u>It is because of these respective media types that ethical and news-related subject matter can sometimes seem different or altered.</u>

a. NO CHANGE
b. It is because of these respective media types, that ethical and news-related subject matter, can sometimes seem different or altered.
c. It is because of these respective media types, that ethical and news-related subject matter can sometimes seem different or altered.
d. It is because of these respective media types that ethical and news-related subject matter can sometimes seem different. Or altered.

40. Which of the following would be the best choice for this sentence (reproduced below)?

(40) These are clearly meant for direct the viewer to the most exciting and important stories the paper has to offer.

a. NO CHANGE
b. These are clearly meant for the purpose of giving direction to the viewer
c. These are clearly meant to direct the viewer
d. These are clearly meant for the viewer to be directed

41. Which of the following would be the best choice for this sentence (reproduced below)?

(41) On a national level there was a story on the evolving Petraeus scandal involving his supposed affair.

a. NO CHANGE
b. On a national level a story was there
c. On a national level; there was a story
d. On a national level, there was a story

42. Which of the following would be the best choice for this sentence (reproduced below)?

Finer details are usually expanded on (42) in written articles, usually people who read newspapers or go online for web articles want more than a quick blurb.

a. NO CHANGE
b. in written articles. People who usually
c. in written articles, usually, people who
d. in written articles usually people who

43. Which of the following would be the best choice for this sentence (reproduced below)?

The diction of the story is also more precise and can be either straightforward or suggestive (43) depending in earnest on the goal of the writer.

a. NO CHANGE
b. depending; in earnest on the goal of the writer.
c. depending, in earnest, on the goal of the writer.
d. the goal of the writer, in earnest, depends on the goal of the writer.

44. Which of the following would be the best choice for this sentence (reproduced below)?

Usually, all news (44) outlets have some sort of bias, it's just a question of how much bias clouds the reporting.

a. NO CHANGE
b. outlets have some sort of bias. Just a question of how much
c. outlets have some sort of bias it can just be a question of how much
d. outlets have some sort of bias, its just a question of how much

45. Which of the following would be the best choice for this sentence (reproduced below)?

(45) <u>However an over use of bias,</u> opinion, and suggestive language can rob readers of the chance to interpret the news events for themselves.

    a. NO CHANGE
    b. However, an over use of bias,
    c. However, with too much bias,
    d. However, an overuse of bias,

Math Test

1. Which of the following inequalities is equivalent to $3 - \frac{1}{2}x \geq 2$?
    a. $x \geq 2$
    b. $x \leq 2$
    c. $x \geq 1$
    d. $x \leq 1$

2. If $g(x) = x^3 - 3x^2 - 2x + 6$ and $f(x) = 2$, then what is $g(f(x))$?
    a. -26
    b. 6
    c. $2x^3 - 6x^2 - 4x + 12$
    d. -2

3. What is the definition of a factor of the number 36?
    a. A number that can be divided by 36 and have no remainder
    b. A number that 36 can be divided by and have no remainder
    c. A prime number that is multiplied times 36
    d. An even number that is multiplied times 36

4. What are the coordinates of the focus of the parabola $y = -9x^2$?
    a. $(-3, 0)$
    b. $\left(-\frac{1}{36}, 0\right)$
    c. $(0, -3)$
    d. $\left(0, -\frac{1}{36}\right)$

5. What is the volume of a cube with the side equal to 3 inches?
    a. 6 in^3
    b. 27 in^3
    c. 9 in^3
    d. 3 in^3

6. What is the volume of a rectangular prism with the height of 3 centimeters, a width of 5 centimeters, and a depth of 11 centimeters?

 a. 19 cm^3

 b. 165 cm^3

 c. 225 cm^3

 d. 150 cm^3

7. What is the volume of a cylinder, in terms of π, with a radius of 5 inches and a height of 10 inches?

 a. 250 π in^3

 b. 50 π in^3

 c. 100 π in^3

 d. 200 π in^3

8. What is the solution to the following system of equations?

$$x^2 - 2x + y = 8$$
$$x - y = -2$$

 a. $(-2, 3)$

 b. There is no solution.

 c. $(-2, 0)\ (1, 3)$

 d. $(-2, 0)\ (3, 5)$

9. An equation for the line passing through the origin and the point $(2, 1)$ is

 a. $y = 2x$

 b. $y = \frac{1}{2}x$

 c. $y = x - 2$

 d. $2y = x + 1$

10. A rectangle was formed out of pipe cleaner. Its length was $\frac{1}{2}$ of a foot and its width was $\frac{11}{2}$ inches. What is its area in square inches?

 a. $\frac{11}{4}$ in^2

 b. $\frac{11}{2}$ in^2

 c. 22 in^2

 d. 33 in^2

11. What type of function is modeled by the values in the following table?

x	$f(x)$
1	2
2	4
3	8
4	16
5	32

a. Linear
b. Exponential
c. Quadratic
d. Cubic

12. Two cards are drawn from a shuffled deck of 52 cards. What's the probability that both cards are Kings if the first card isn't replaced after it's drawn?

a. $\frac{1}{169}$

b. $\frac{1}{221}$

c. $\frac{1}{13}$

d. $\frac{4}{13}$

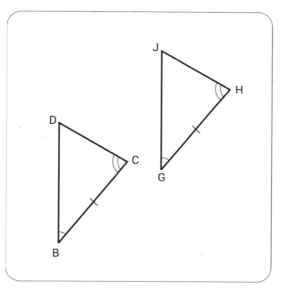

13. In the image above, what is demonstrated by the two triangles?
a. According to Side-Side-Side, the triangles are congruent.
b. According to Angle-Angle-Angle, the triangles are congruent.
c. According to Angle-Side-Angle, the triangles are congruent.
d. There is not enough information to prove the two triangles are congruent.

14. Which expression is the same as six less than three times the sum of twice a number and one?
   a. $2x + 1 - 6$
   b. $3x + 1 - 6$
   c. $3(x + 1) - 6$
   d. $3(2x + 1) - 6$

15. If $\sqrt{5 + x} = 5$, what is x?
   a. 10
   b. 15
   c. 20
   d. 25

16. Which of the following polynomials is equal to $(2x - 4y)^2$?
   a. $4x^2 - 16xy + 16y^2$
   b. $4x^2 - 8xy + 16y^2$
   c. $4x^2 - 16xy - 16y^2$
   d. $2x^2 - 8xy + 8y^2$

17. What are the zeros of $f(x) = x^2 + 4$?
   a. $x = -4$
   b. $x = \pm 2i$
   c. $x = \pm 2$
   d. $x = \pm 4i$

18. Which of the following shows the correct result of simplifying the following expression:

$$(7n + 3n^3 + 3) + (8n + 5n^3 + 2n^4)$$

   a. $9n^4 + 15n - 2$
   b. $2n^4 + 5n^3 + 15n - 2$
   c. $9n^4 + 8n^3 + 15n$
   d. $2n^4 + 8n^3 + 15n + 3$

19. What is the simplified result of $\frac{15}{23} \times \frac{54}{127}$?
   a. $\frac{810}{2,921}$
   b. $\frac{81}{292}$
   c. $\frac{69}{150}$
   d. $\frac{810}{2929}$

20. What is the product of the following expression?

$$(4x - 8)(5x^2 + x + 6)$$

    a. $20x^3 - 36x^2 + 16x - 48$
    b. $6x^3 - 41x^2 + 12x + 15$
    c. $204 + 11x^2 - 37x - 12$
    d. $2x^3 - 11x^2 - 32x + 20$

21. What is the solution for the following equation?

$$\frac{x^2 + x - 30}{x - 5} = 11$$

    a. $x = -6$
    b. There is no solution.
    c. $x = 16$
    d. $x = 5$

22. If x is not zero, then $\frac{3}{x} + \frac{5u}{2x} - \frac{u}{4} =$

    a. $\dfrac{12 + 10u - ux}{4x}$

    b. $\dfrac{3 + 5u - ux}{x}$

    c. $\dfrac{12x + 10u + ux}{4x}$

    d. $\dfrac{12 + 10u - u}{4x}$

23. What are the zeros of the function: $f(x) = x^3 + 4x^2 + 4x$?
    a. -2
    b. 0, -2
    c. 2
    d. 0, 2

24. Is the following function even, odd, neither, or both?

$$y = \frac{1}{2}x^4 + 2x^2 - 6$$

    a. Even
    b. Odd
    c. Neither
    d. Both

25. Which of the following formulas would correctly calculate the perimeter of a legal-sized piece of paper that is 14 inches long and $8\frac{1}{2}$ inches wide?

    a. $P = 14 + 8\frac{1}{2}$

    b. $P = 14 + 8\frac{1}{2} + 14 + 8\frac{1}{2}$

    c. $P = 14 \times 8\frac{1}{2}$

    d. $P = 14 \times \frac{17}{2}$

26. A grocery store is selling individual bottles of water, and each bottle contains 750 milliliters of water. If 12 bottles are purchased, what conversion will correctly determine how many liters that customer will take home?

    a. 100 milliliters equals 1 liter
    b. 1,000 milliliters equals 1 liter
    c. 1,000 liters equals 1 milliliter
    d. 10 liters equals 1 milliliter

27. Given the following triangle, what's the length of the missing side? Round the answer to the nearest tenth.

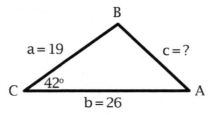

    a. 17.0
    b. 17.4
    c. 18.0
    d. 18.4

28. For the following similar triangles, what are the values of x and y (rounded to one decimal place)?

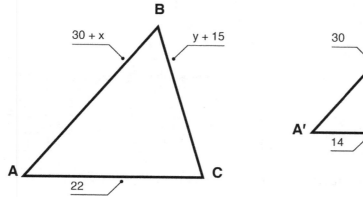
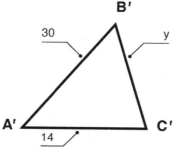

    a. $x = 16.5, y = 25.1$

    b. $x = 19.5, y = 24.1$

    c. $x = 17.1, y = 26.3$

    d. $x = 26.3, y = 17.1$

29. What are the center and radius of a circle with equation $4x^2 + 4y^2 - 16x - 24y + 51 = 0$?

    a. Center $(3, 2)$ and radius $1/2$

    b. Center $(2, 3)$ and radius $1/2$

    c. Center $(3, 2)$ and radius $1/4$

    d. Center $(2, 3)$ and radius $1/4$

30. What is the solution to $(2 \times 20) \div (7 + 1) + (6 \times 0.01) + (4 \times 0.001)$?

    a. 5.064

    b. 5.64

    c. 5.0064

    d. 48.064

31. A piggy bank contains 12 dollars' worth of nickels. A nickel weighs 5 grams, and the empty piggy bank weighs 1050 grams. What is the total weight of the full piggy bank?

    a. 1,110 grams

    b. 1,200 grams

    c. 2,250 grams

    d. 2,200 grams

32. Last year, the New York City area received approximately $27\frac{3}{4}$ inches of snow. The Denver area received approximately 3 times as much snow as New York City. How much snow fell in Denver?

    a. 60 inches

    b. $27\frac{1}{4}$ inches

    c. $9\frac{1}{4}$ inches

    d. $83\frac{1}{4}$ inches

33. If $-3(x + 4) \geq x + 8$, what is the value of x?
   a. $x = 4$
   b. $x \geq 2$
   c. $x \geq -5$
   d. $x \leq -5$

34. Karen gets paid a weekly salary and a commission for every sale that she makes. The table below shows the number of sales and her pay for different weeks.

Sales	2	7	4	8
Pay	$380	$580	$460	$620

Which of the following equations represents Karen's weekly pay?
   a. $y = 90x + 200$
   b. $y = 90x - 200$
   c. $y = 40x + 300$
   d. $y = 40x - 300$

35. The square and circle have the same center. The circle has a radius of r. What is the area of the shaded region?

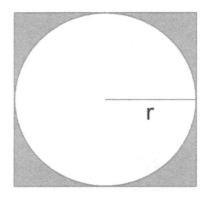

   a. $r^2 - \pi r^2$
   b. $4r^2 - 2\pi r$
   c. $(4 - \pi)r^2$
   d. $(\pi - 1)r^2$

36. The graph shows the position of a car over a 10-second time interval. Which of the following is the correct interpretation of the graph for the interval 1 to 3 seconds?

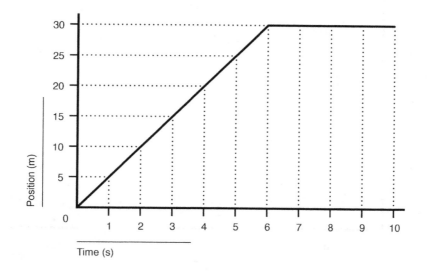

Time (s)

a. The car remains in the same position.
b. The car is traveling at a speed of 5 m/s.
c. The car is traveling up a hill.
d. The car is traveling at 5 mph.

37. Which of the ordered pairs below is a solution to the following system of inequalities?

$$y > 2x - 3$$
$$y < -4x + 8$$

a. (4, 5)
b. (-3, -2)
c. (3, -1)
d. (5, 2)

38. Which equation best represents the scatterplot below?

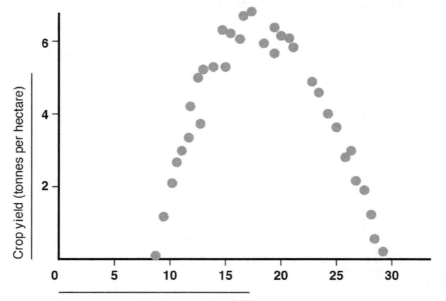

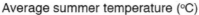

a. $y = 3x - 4$
b. $y = 2x^2 + 7x - 9$
c. $y = (3)(4^x)$
d. $y = -\frac{1}{14}x^2 + 2x - 8$

39. What is the solution to $9 \times 9 \div 9 + 9 - 9 \div 9$?
   a. 0
   b. 17
   c. 81
   d. 9

40. What is the solution to the radical equation $\sqrt[3]{2x + 11} + 9 = 12$?
   a. -8
   b. 8
   c. 0
   d. 12

41. The hospital has a nurse to patient ratio of 1:25. If there is a maximum of 325 patients admitted at a time, how many nurses are there?
   a. 13 nurses
   b. 25 nurses
   c. 325 nurses
   d. 12 nurses

42. A hospital has a bed to room ratio of 2 to 1. If there are 145 rooms, how many beds are there?
    a. 145 beds
    b. 2 beds
    c. 90 beds
    d. 290 beds

43. If $\frac{2x}{5} - 1 = 59$, what is the value of x?
    a. 60
    b. 145
    c. 150
    d. 115

44. A National Hockey League store in the state of Michigan advertises 50% off all items. Sales tax in Michigan is 6%. How much would a hat originally priced at $32.99 and a jersey originally priced at $64.99 cost during this sale? Round to the nearest penny.
    a. $97.98
    b. $103.86
    c. $51.93
    d. $48.99

45. Store brand coffee beans cost $1.23 per pound. A local coffee bean roaster charges $1.98 per 1 and a half pounds. How much more would 5 pounds from the local roaster cost than 5 pounds of the store brand?
    a. $0.55
    b. $1.55
    c. $1.45
    d. $0.45

46. Paint Inc. charges $2000 for painting the first 1,800 feet of trim on a house and $1.00 per foot for each foot after. How much would it cost to paint a house with 3125 feet of trim?
    a. $3125
    b. $2000
    c. $5125
    d. $3325

No Calculator Questions

47. A bucket can hold 11.4 liters of water. A kiddie pool needs 35 gallons of water to be full. How many times will the bucket need to be filled to fill the kiddie pool?
    a. 12
    b. 35
    c. 11
    d. 45

48. In Jim's school, there are 3 girls for every 2 boys. There are 650 students in total. Using this information, how many students are girls?

    a. 260
    b. 130
    c. 65
    d. 390

49. What is the volume of a pyramid, with a square base whose side is 6 inches, and the height is 9 inches?

    a. 324 in^3
    b. 72 in^3
    c. 108 in^3
    d. 18 in^3

50. Convert $\frac{2}{9}$ to a percentage.

    a. 22%
    b. 4.5%
    c. 450%
    d. 0.22%

51. What is the volume of a cone, in terms of π, with a radius of 10 centimeters and height of 12 centimeters?

    a. 400 cm^3
    b. 200 cm^3
    c. 120 cm^3
    d. 140 cm^3

52. What is 3 out of 8 expressed as a percent?

    a. 37.5%
    b. 37%
    c. 26.7%
    d. 2.67%

53. The area of a given rectangle is 24 square centimeters. If the measure of each side is multiplied by 3, what is the area of the new figure?

    a. 48 cm^2
    b. 72 cm^2
    c. 216 cm^2
    d. 13,824 cm^2

54. If $4x - 3 = 5$, what is the value of x?

55. What is the solution to $4 \times 7 + (25 - 21)^2 \div 2$?

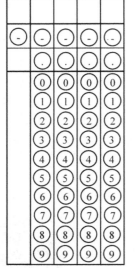

56. What is the solution to the following expression?

$$(\sqrt{36} \times \sqrt{16}) - 3^2$$

57. What is the overall median of Dwayne's current test scores: 78, 92, 83, 97?

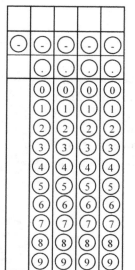

58. The total perimeter of a rectangle is 36 cm. If the length is 12 cm, what is the width?

Essay Prompt

For-profit institutions of learning should be illegal because they suffer from a conflict of interest between the student receiving the best education and the institution minimizing costs in providing instruction and other services to remain profitable.

Write a response in support or nonsupport of the statement in regard to the position that is taken. To support your stance, evaluate how your position may or may not be true to the argument. Explain how you've come to your position and provide examples in support.

Answer Explanations for Practice Test #1

Reading Test

1. C: There is a tone of excitement throughout the passage; this tone is conveyed through the narrator's account of the life-changing discovery of electricity. Mentions of tunnels and darkness could evoke a tone of dreariness, but that represents the past. This passage is looking forward to what the future will hold with this new form of light, making Choice A incorrect. Choice B is incorrect because the narrator does not feel unnerved when about the idea of confessing their crime to the Council. This passage does not use humor or lightheartedness, making Choice D incorrect.

2. A: Evidence for the correct answer can be found here: "They would see nothing, save our crime of working alone, and they would destroy us and our light." In this society, things must be done as a collective, not individually. The narrator hopes to join the Home of the Scholars, and wants their input and work space, which makes Choice B incorrect. Street sweeping is the narrator's role but it's not what makes their discovery a crime. That is why Choice C is incorrect. Besides the word *Unmentionable*, nothing indicates that remainders from former times are off-limits, making Choice D incorrect.

3. D: The sentence uses personification, whereby darkness is given a human quality—being able to swallow. There are no comparison being made, making metaphor, Choice A, incorrect. Synecdoche, which is a device that uses a part to represent a whole or vice versa is not present, making Choice B incorrect. The sentence has no interjected scenes from a previous point in the narrative, making flashback, Choice C, incorrect.

4. A: Multiple times the narrator calls their discovery "the power of the sky" which is an allusion to the sun. For example, in the third paragraph, the narrator reports, "we finished building a strange thing…a box of glass, devised to give forth the power of the sky of greater strength than we had ever achieved before." Candles were a source of light before this discovery, making Choice B incorrect. Prison represents the darkness, not the light, making Choice C incorrect. The narrator does feel that their invention is a part of their body, the way blood runs through their veins, but does not directly equate to their overall invention of electricity. This makes Choice D incorrect.

5. B: The narrator is confident their crime will be forgiven by the World Council of Scholars, because their "gift is greater than [their] transgression." The passage does not address the money their invention will save the City, making Choice A incorrect. It is not the apology alone that will grant the narrator forgiveness, making Choice C incorrect. By the end, the narrator does feel a strong connection to their invention, but the wires are only metaphorically part of their body, and the wires are not the reason the narrator believes they'll be forgiven. This makes Choice D incorrect.

6. D: *Transgression* most closely means *offense* in paragraph eight. Obedience, or following the rules, is the opposite of transgression, making Choice A incorrect. The word *transgression* has a moral or legal connotation that is lacking in the more abstract word *disruption*, making Choice B incorrect. Confession, or admitting to an offense, is what the narrator is going to do as a result of their transgression, making Choice D incorrect.

7. B: Personification is a literary device in which human characteristics are attributed to something nonhuman. So when the narrator says "darkness swallowed us," they give a human characteristic (swallowing) to something nonhuman (darkness). Choice A is incorrect because it is just an expression used to indicate there are a lot of thoughts swirling around in the narrator's head. Choice *C* is refers to the human narrator's head and hands; since the narrator is a person, they cannot be personified. Choice *D* is incorrect because it is a thought that trails off using ellipsis, not personification, to indicate that the narrator has changed their opinion on the value of their body.

8. C: The narrator wants to keep their invention a secret until they are able to confess everything to the World Council of Scholars. Paragraph seven lists the resources that are necessary to advance the discovery. Those include work rooms, which makes Choice *A* incorrect. The narrator also wants help and wisdom from the brother Scholars, making Choice *B* incorrect. The narrator feels that they will need to focus all their time on their discovery, which means they will need to stop sweeping the streets, leaving Choice *D* incorrect.

9. A: Moving from street sweeper to the Home of the Scholars will be an advancement in society. This is apparent because the narrator currently lives in a tunnel with no electricity, while the wisest in all the lands are elected to the World Council of Scholars. Choice *B* is incorrect because the text indicates that being a Scholar is more desirable than being a street sweeper, so their status will not remain the same. The Council of Vocations decides what jobs people will have, and would be the group to assign the narrator to the House of Scholars. This makes Choice *C* incorrect. The narrator is not concerned with the negative consequences of their confession and is confident they will be propelled to Scholar status. This is why Choice *D* is incorrect.

10. D: The last paragraph reveals the shift in how the narrator suddenly cares about their body for the first time after they discover electricity. The passage cuts off before the author shares whether the transgression is forgiven, making Choice *A* incorrect. The plan is to share the gift of electricity with all the Cities, but the reader does not know if that plan will come to fruition. This makes Choice *B* incorrect. It's likely that electricity will make all tasks, especially those that happen at night, easier. But Choice *C* is incorrect because those benefits are not discussed in this passage.

11. C: Jimmy Carter talks about how Americans are losing confidence in their government, their ability to self-govern, and democracy. He basically says America has lost its way and must find it again. That will be hard and will require facing the truth. In the last paragraph, he says that we need to change course and restore confidence and faith. Choice *A* is incorrect because Carter believes that faith can be restored, meaning there is hope for the future. Choice *B* only addresses part of the speech, so it is not the best answer. While Carter does reference other people in the Western world, that is irrelevant to the underlying message, making Choice *D* incorrect.

12. B: Worshiping self-indulgence and consumption is cited as a symptom of the crisis of the American spirit. Additionally, Choices *C* and *D* are both listed in the third paragraph as examples of symptoms of the crisis of the American spirit, which makes them incorrect answers. Another symptom is that the majority of people believe the next five years will be worse than the last five. However, Choice *B* is wrong. Carter says, "there is a growing disrespect for government and for churches and for schools, the news media, and other institutions." Therefore, respect isn't growing, *disrespect* is,

13. B: Using context clues from the passage, a reader can infer that the saying *sound as a dollar* means something is dependable and reliable. *Solid as a rock* is another phrase that means dependability and reliability. *Hands down* refers to intensity or completeness; something that is "the best value, hands down," is absolutely the best value. The phrase *hands down* does not speak to dependability, making Choice *A* incorrect. Ease, not reliability, is the meaning behind the phrase *piece of cake*. This makes

Choice C incorrect. Choice D is incorrect because getting something fair and square means it was earned honestly.

14. C: The goal of Carter's speech is to sound relatable and capable of understanding of Americans' frustration with government. Carter emphasizes his empathy when he says, "Often you see paralysis and stagnation and drift. You don't like it, and neither do I. What can we do?" Nothing in his speech would indicate he is unqualified for the role of president, making Choice A incorrect. Carter may be desperate to restore faith and confidence in Americans, but not desperate for their acceptance. This makes Choice B incorrect. Choice D is incorrect because Jimmy Carter fully believes in democracy and is encouraging Americans to believe in it.

15. D: Out of context, Washington, D.C. being an island could mean many things. The meaning of the metaphor can be found in the sentences that precede and follow it, which mention isolation and a wide gap between the government and its citizens. Choice A is wrong because Carter is not mentioning an island to describe a vacation. This metaphor does not allude to resource, so Choice B is incorrect. Choice C implies that Americans have access to the "island" of government, but the real meaning is the opposite, that government has become unresponsive to citizens. Therefore, Choice C is incorrect.

16. B: According to Carter, these examples are the shocks and tragedies that have gradually caused Americans to lose their faith and confidence in government and other institutions such as schools and media. While there are references to previous presidents, it is not Carter's intention to grade their time in office, making Choice A incorrect. The future is not mentioned in paragraphs six and seven, rendering Choice C incorrect. President Carter believes Americans have very reasonable expectations, even when they're not being met. For that reason, Choice D is incorrect.

17. C: It is in paragraph eight that Carter mentions *politics as usual,* and then describes it in detail in paragraph nine. He does it at the same time he is describing the negative aspects of politics, like false claims, evasiveness, inactivity, and a lack of support. Carter does not mention how decisions affect any of the branches of government, making Choice A wrong. Choice B is incorrect because there is no textual evidence to support Carter's concern with how political language reads to the average American. According to Carter, a balanced and fair approach that demands sacrifice is what politics should look like, but is rarely, if ever, the case. This is what makes Choice D incorrect.

18. A: Stagnation is a lack of movement; in this context, it is a lack of forward movement or progress toward ideals. Context clues like "incapable of action" and "paralysis" help define stagnation. *Confusion* would be a better word for lack of understanding, which makes Choice B incorrect. *Ignorance* would be a better word for lack of knowledge, making Choice C incorrect. A lack of empathy would be apathy, not stagnation, making Choice D wrong as well.

19. C: President Carter begins by addressing the events that have led to the crisis of confidence in America, such as the murder of Martin Luther King Jr. and the inflation and collapse of the American dollar. The last paragraph offers a message of hope for the future, ending in an optimistic tone. Since this is a presidential speech, the entire speech will be formal, making Choice A incorrect. The speech remains elevated and the last paragraph offers hope for the future, the opposite of unremarkable. This makes Choice B incorrect. Since President Carter wants to be relatable and empathetic, he avoids sounding preachy, which is why Choice D is wrong.

20. C: In the final paragraph, Carter answers the question of what to do; he gives a plan to face the truth and restore faith. There are wounds from the past, but Carter does not suggest healing them as a solution to the crisis. This makes Choice A incorrect. Choice B is incorrect because Carter admits special interest groups are detrimental to running a smooth government but does not propose removing them to mend the crisis. Close-knit communities are a healthy side effect of, not the solution to, restoring faith and confidence in America. This is why Choice *D* is incorrect.

21. A: Truman presented his argument to Congress in order to persuade it to take action to support Greece and Turkey. Truman does not rely on descriptive language, so Choice *B* is not accurate. This speech did not intend to entertain listeners, since there are no attempts at humor or amusement, making Choice *C* incorrect. This speech does analyze a situation, but analysis is not the main purpose of the speech, making Choice *D* wrong.

22. D: When Truman says, "I believe...I believe...I believe...," he means the United States is obligated to take care of other countries in need. Truman is concerned Greece and Turkey will fall under the sway of a small number of armed insurgents. Choice *A* is wrong because there is not any information in the passage that deals with the economic status of Greece and Turkey Choice *B* is incorrect because Truman doesn't list the countries that are in the United Nations. The passage does not describe anything about whether Greece nor Turkey is under a dictatorship, so we cannot infer this information from what is provided. This makes Choice *C* incorrect.

23. C: In the first way of life, countries ruled by a majority, with representative government, free elections, and freedom of speech. In the second way of life, the will of a small number of people is forcibly imposed upon the majority. These countries are characterized by terror and oppression. Choice *A* is wrong because both parts of the sentence describe the first way of life. Choice *B* is wrong because both qualities describe the second way of life. Choice *D* is incorrect because it is the opposite of what Truman describes; the first part describes the second way of life, and vice versa.

24. B: The word *subterfuges* could be replaced with *tricky deceptions* and the sentence would retain the same meaning. This can be gathered from nearby words like *violation, coercion*, and *infiltration*. *Helpful pieces of intel* would be information of military or political value, not deceit, which makes Choice *A* wrong. Political infiltration may be a time-consuming activity, but it does not address the deception of subterfuge. This leaves Choice *C* incorrect. The prefix "sub-" in the word *subterfuge* does mean "under," but it would only mean underground in a word like *subterranean*. This makes Choice *D* wrong.

25. A: President Truman argues that without interference from the United States, Greece will inevitably be oppressed by a small number of people who take up arms against the majority of the people. That will cause a ripple through Turkey, the entire Middle East, and beyond. Nothing indicates that the United Nations would step in, nor is it the responsibility of the UN to provide relief funds. This makes Choice *B* incorrect. The Truman Doctrine is a plan of preventative measures, to help Greece and Turkey recover from war. There is not a logical timeline where Truman's plan is not approved by Congress, Greece falls under the control of an armed minority, and then the United States retracts and sends hundreds of millions of dollars to Greece. This makes Choice *C* invalid. Choice *D* is incorrect because the only mention of Greece disappearing was in reference to its status as a free state.

26. C: Truman's warning about what will happen if the United States fails to help Greece and Turkey should leave his listeners with a sense of concern. Congress would feel disgusted if Truman described something revolting. This is why Choice *A* is incorrect. Choice *B* is incorrect because an audience would feel astonishment if Truman shared surprising new information. Choice *D* is not the best option because there are other sentences from the speech that would be more likely to evoke an eagerness to help (e.g., "This is a serious course upon which we embark.").

27. B: Nowhere does Truman mention overhauling the governments of Greece and Turkey. He actually believes Greece and Turkey can help themselves when they are provided with ample funds and resources. Choices *A, C,* and *D* can all be found directly in paragraphs fourteen, fifteen, and sixteen, which is what makes them incorrect answers.

28. C: President Truman felt the United States was the sole entity responsible for helping free people resisting a takeover from armed aggressors or outside threats. Truman does not allude to the possibility of another country stepping in, which is why Choice *A* is doubtful and incorrect. Since Greece and Turkey will both be in similar states of turmoil it is unlikely they will be of any help to one another. This makes Choice *B* wrong. It is not the responsibility of the United Nations to intervene in these matters, and cannot force the United States to send funds to foreign countries, which renders Choice D incorrect.

29. D: The use of the phrase "immediate and resolute action" conveys a sense of urgency. Choice *A* is about the cost of safeguarding freedom; it does not refer to time, which makes it incorrect. Choice *B* focuses on how the world is always changing, but it does not mention time, making it wrong as well. Choice *C* is incorrect because while it does speak to the gravity of the situation, it does not attest to the time-sensitive nature of the matter.

30. A: The vulnerability Greece and Turkey is the problem, and the proposed three-step process that involves sharing funds and resources is the solution. Since there is only one date listed and events are not structured in a timeline, this speech is not organized chronologically. That makes Choice *B* incorrect. Other than a quick comparison between two ways of life, Truman's speech does not rely on comparing and contrasting, making Choice *C* incorrect. Categorical structure usually involves the classification of equally important groups or subgroups, which is not observed here, making Choice *D* incorrect.

31. B: Although these passages are written in completely different styles, they're both centered around the role insects play in pollination. Choice *A* is incorrect because only the second article breaks down the role of the three different sets of legs. The first article is not written scientifically, making Choice *C* wrong. Only the first passage goes into depth about nectar and the signals plants use to communicate to bees. This is why Choice *D* is not the best answer.

32. C: The passage by Jean-Henri Fabre was written for a children's storybook, which is why he uses whimsical language. Personifying both the bee and the flower and using terms like "signboard," "enter here," and "keyhole" evoke the mood of a fairytale or fable. Pedantic word choice suggests a pompous scholarly paper. This passage is anything but pompous and scholarly, making Choice *A* incorrect. This passage is light-hearted, while something emotional usually plays on deep or dark emotions. This is why Choice *B* is incorrect. Choice *D* is incorrect because, despite whimsical storytelling, the message and information are clear, not ambiguous.

33. D: The best substitution for the word *posterior* is *rear*. This answer can be drawn from realizing posterior legs means the opposite of "forward appendages." Also the prefix *post-* means behind or after. While the word *strongest* fits in the sentence nicely, nothing implies that that particular set of legs is the strongest. This makes Choice *A* incorrect. *Anterior*, meaning front, is actually the antonym for the word *posterior*, so Choice *B* is wrong. *Ambulatory* is a word that could describe any pair of functioning legs and does not differentiate the hind legs from the front or middle, which makes Choice *C* incorrect.

34. D: The scientists in the second passage use observation to conduct research on honeybees. This is set up at the beginning of the passage when the author mentions continuous observation and instantaneous photos. The author does not use the art of storytelling to reinforce their research, making the word *anecdotal*, Choice *A*, wrong. The researcher collected data by observing the pollination process between bees and sweet corn, none of which was hypothetical. This makes Choice *B* incorrect. Choice *C* is incorrect because the researchers did not conduct tests or introduce variables, meaning the evidence was merely observational, not experimental.

35. A: The character believes that the snapdragon's yellow spot is evidence that Providence or some higher power has designed nature. When a bee lands on the yellow spot of a closed snapdragon, the flower opens up so the bee can access the nectar and pollen. The preceding sentence confirms that the character does not believe that nature happens by chance, making Choice *B* incorrect. The character makes no claim that snapdragons are more beautiful than any other plant. This makes Choice *C* wrong. This character's speech continuously highlights the intentional behaviors between flowers and bees, so Choice *D* is incorrect.

36. C: The triple feast, introduced in the first paragraph of the first passage, has three beneficiaries: trees, insects, and humans. Pollen is part of the transaction between these three, but pollen not directly benefit, which is why Choices *A*, *B*, and *D* are incorrect.

37. B: When bees interact with the anthers of the flowers, the hairs of their bodies get covered in pollen. The bee's legs collect the pollen from all over the body and transfer it back toward the pollen baskets, also known as the corbiculae. Choice *B* best explains the order of the pollen manipulation process. Choices *A*, *C*, and *D* incorrectly describe (either by order or parts) the way bees move pollen across their bodies.

38. C: Synonyms are two words that have the same meaning. The second-to-last paragraph is where the reader learns that *corbiculae* is another term for pollen baskets. Nectar is what bees find inside the flowers, and honey is a result of what bees do with the nectar. This is why Choice *A* is incorrect. Stamen and stigma are both parts of the anatomy of a flower but serve different purposes, which makes Choice *B* incorrect. Choice *D* is wrong because the thorax is adjacent to the abdomen; it is not the same part of the bee's body.

39. D: In a false dichotomy, also known as an either-or fallacy, only two alternative points are presented, even though other alternative points are available. The author is indirectly saying that if someone does not understand the relationship between the bee and the snapdragon, they are an idiot. That is a rather harsh two-option opinion. A slippery slope fallacy is when someone claims that a series of small events will necessarily result in one major bad event, which is not presented here, making Choice *A* incorrect. Choice *B*, ad hominem, is incorrect because the author does not attack another person's character. There are no distractions or misleading clues that would indicate a red herring fallacy, making Choice *C* wrong.

40. A: The author means the flower's petals open up slowly like a person yawning. "To open" or "to gape" is one of the meanings of the verb "yawn." Choice *B* is incorrect because it focuses on the sleepiness of yawning, not the motion. Choice *C* is wrong because it inaccurately parallels how yawning allows people to receive more oxygen to their brains. There is no evidence to indicate the flower is having issues with regulation or time of day, making Choice *D* erroneous.

41. D: The researchers chose sweet corn because sweet corn produces an abundance of pollen, making it easier to observe. The passage does not mention that bees are more attracted to sweeter flowers or that sweet corn is particularly sweet to bees. This makes Choice *A* incorrect. Sweet corn was used because of its abundance of pollen, not its abundance of itself, making Choice *B* incorrect. While the article does focus on bees, it does not address other pollinators, leaving Choice *C* incorrect.

42. B: Although written in very different ways, both passages describe a mutually beneficial relationship between bees and flowers when it comes to pollination. Based on the level of detail provided, it is safe to say that the behavior of honeybees and pollen collection is well studied, making Choice *A* incorrect. Even though both passages focus their attention on honeybees, neither one states that bees are superior pollinators, making Choice *C* wrong. Both authors feel very strongly that the act of pollination is purposeful and intentional, and they feel that nothing can be attributed to randomness. This is what makes Choice *D* incorrect.

43. C: *Solution and Emulsions* is not just the correct answer, it is the exact title of the passage from *Common Science*. The title *Solutions and Emulsions* fully covers what this excerpt is about. Choice *A* is incorrect because it too narrowly focuses on the aspect of cleaning when that is only a side benefit to learning about solutions and emulsions. Choice *B* is wrong because this passage did not contain experiments about any topic other than solutions and emulsions. Choice *D* is overly focused on the opening and doesn't account for the rest of the passage, making it incorrect.

44. A: The author conjures up a world where there is an imaginary switchboard that can turn off the law of solutions, runs through an example where all saltwater fish die, and then decides the switchboard should be left alone. This is a comedic hypothetical to open the topic of solutions and emulsions. There is not an element of shock or surprise, making Choice *B* incorrect. Despite the fiction, all the information presented was logical and complete, which leaves Choice *C* incorrect. Since the author tells the reader exactly what happens when the Solutions switch is turned off, there is no element of suspense, making Choice *D* incorrect.

45. B: All of the rhetorical questions are designed to make the reader think about the results and effects of the different experiments, eventually leading them to draw their own conclusions. Science is made up of facts, not opinions, so Choice *A* is incorrect. These rhetorical questions have a greater purpose than to just ensure the readers are following along. This makes Choice *C* incorrect. There is an assumption the reader is already invested in doing the experiments and would not need prodding from rhetorical questions, which is why Choice *D* is wrong.

46. D: Evidence for this answer can be derived from the example of milk and its emulsion tendencies. Little droplets of butter and fat are suspended in milk and are not completely dissolved like sugar would be in water when making simple syrup. Since sugar can completely dissolve into water, simple syrup is a solution. Choice *A* is incorrect because solutions are actually more likely to be clear. Both emulsions and solutions can be drinkable or non-drinkable, just depends on what components are being used. This makes Choice *B* incorrect. Choice *C* is erroneous because emulsions and solutions do not naturally have lower or higher temperatures.

47. B: After the experimenter is asked to add more salt than can be dissolved and then to use a Bunsen burner to heat the test tube, a logical result would be that the heat is an agent to dissolve the salt more readily. Choice *A* is the opposite of what result should be expected. The rhetorical question that follows the instructions implies that the salt will not impede the heating element, making Choice *C* incorrect. While the author does mention washing dishes, using the saltwater solution as soap is an inaccurate connection, leaving Choice *D* incorrect.

48. A: The passage mentions briefly that emulsions can act as solutions if their suspended droplets are so small they can't be identified. Heating can help dissolve more solute into solvent but is not the reason emulsions can act like solutions, making Choice *B* incorrect. Shaking and stirring does not change the smallest size droplets can be inside an emulsion, rendering Choice *C* incorrect. Because the passage directly says "But the droplets may be so small that an emulsion acts almost exactly like a solution," Choice *D* cannot be correct.

49. B: The last paragraph offers an analogy to better explain the results of Experiment 82. When a pail of pebbles is poured into a pail of apples, the pebbles will fill in the empty spaces between the apples. Likewise, when water is poured into alcohol, the water molecules will rest between the molecules of alcohol, meaning the combined liquids of equal measure do not quite double in size. Choice *A* and *C* both get the analogy backwards. "Pebbles are to apples" is a statement of how they relate; the pebbles fill the spaces between the apples. The second half should therefore be "as water is to alcohol." Therefore Choice *A* is incorrect. Choice *C* also incorrectly equates the apples to water molecules and the pebbles to alcohol molecules. Choice *D* is incorrect because the scenario of the pebbles and the apples is meant to explain emulsion; the apples and pebbles are not in solution, and neither are the alcohol and the water.

50. D: Even if you do not know that denatured means the alcohol has been altered in some fashion, there are context clues that imply the solution will not be pure due to the denatured substances. There is no evidence to indicate denatured alcohol is dangerous or unfit for a laboratory, making Choice *A* wrong. Choice *B* is incorrect because the author wrote an entire paragraph differentiating denatured alcohol from regular alcohol. And there is no amount of heating that will cause the denatured components to dissolve, which is what makes Choice *C* incorrect.

51. C: The apple and pebble analogy in the last paragraph explains the purpose of Experiment 82. Because the water molecules will fill the gaps between the larger alcohol molecules, it can be deduced that the mixture will measure less than two inches, which is the combined separate measurements of water and alcohol. The experiment would not prove anything if the total height measured exactly two inches, making Choice *A* incorrect. Nothing in the passage indicates the alcohol or water would expand when combined, leaving Choice *B* incorrect. There is enough surrounding evidence to make a reasonable prediction about the result of Experiment 82, which makes Choice *D* wrong.

52. C: When two people are described as oil and water it means they do not mix well or they are incompatible. This passage explains that oil and water can only be an emulsion because the two substances can never fully combine into a homogenous solution. The idiom *comparing apples to oranges* means you are comparing two unlike things, which is not explained in this passage. That makes Choice *A* incorrect. When someone is *in hot water* it means they are in trouble, which is not described in this passage, making Choice *B* wrong. When someone says *do not cry over spilled milk*, they mean advising someone not be upset over something that happened in the past and cannot be change. Choice *D* is incorrect because we learned that milk is an emulsion, but there was no advice about what to do when it is spilled.

Writing and Language Test

1. C: Choice *C* correctly uses *from* to describe the fact that dogs are related to wolves. The word *through* is incorrectly used here, so Choice *A* is incorrect. Choice *B* makes no sense. Choice *D* unnecessarily changes the verb tense in addition to incorrectly using *through*.

2. B: Choice *B* is correct because the Oxford comma is applied, clearly separating the specific terms. Choice *A* lacks this clarity. Choice *C* is correct but too wordy since commas can be easily applied. Choice *D* doesn't flow with the sentence's structure.

3. D: Choice *D* correctly uses the question mark and includes the verb, fixing the sentence's main issues. Thus, Choice *A* is incorrect because questions do not end with periods. Choice *B*, although correctly written, changes the meaning of the original sentence. Choice *C* is incorrect because it completely changes the direction of the sentence, disrupts the flow of the paragraph, and lacks the crucial question mark.

4. A: Choice *A* is correct since there are no errors in the sentence. Choices *B* and *C* both have extraneous commas, disrupting the flow of the sentence. Choice *D* unnecessarily rearranges the sentence.

5. D: Choice *D* is correct because the commas serve to distinguish that *artificial selection* is just another term for *selective breeding* before the sentence continues. The structure is preserved, and the sentence can flow with more clarity. Choice *A* is incorrect because the sentence needs commas to avoid being a run-on. Choice *B* is close but still lacks the required comma after *selection*, so this is incorrect. Choice *C* is incorrect because the comma to set off the aside should be placed after *breeding* instead of *called*.

6. B: Choice *B* is correct because the sentence is talking about a continuing process. Therefore, the best modification is to add the word *to* in front of *increase*. Choice *A* is incorrect because this modifier is missing. Choice *C* is incorrect because, with the additional comma, the present tense of *increase* is inappropriate. Choice *D* makes more sense, but the tense is still not the best to use.

7. A: The sentence has no errors, so Choice *A* is correct. Choice *B* is incorrect because it adds an unnecessary comma. Choice *C* is incorrect because *advantage* should not be plural in this sentence unless the singular article *an* is removed. Choice *D* is very tempting. Although this would make the sentence more concise, it leaves out critical information, which makes it incorrect.

8. C: Choice *C* correctly uses *on to*, describing the way genes are passed generationally. The use of *into* is inappropriate for this context, which makes Choice *A* incorrect. Choice *B* is close, but *onto* refers to something being placed on a surface. Choice *D* doesn't make logical sense.

9. D: Choice *D* is correct, since only proper names should be capitalized. Because "bloodhound" is not a proper name, Choice *A* is incorrect. Only the proper nouns within breed names need to be capitalized, such as a German shepherd. In terms of punctuation, only one comma after example is needed, so Choices *B* and *C* are incorrect.

10. D: Choice *D* is the correct answer because *rather* acts as an interrupting word here and thus should be separated by commas. Choices *B* and *C* use commas unwisely, breaking the flow of the sentence.

11. B: Since the sentence can stand on its own without *usually*, separating it from the rest of the sentence with a comma is correct. Choice *A* needs the comma after *usually*, while Choice *C* uses commas incorrectly. Choice *D* is tempting, but changing *turn* to past tense goes against the rest of the paragraph.

12. A: In Choice *A*, the dependent clause "Sometimes in particularly dull seminars" is seamlessly attached with a single comma after *seminars*. Choice *B* contains too many commas. Choice *C* does not correctly combine the dependent clause with the independent clause. Choice *D* introduces too many unnecessary commas.

13. D: Choice *D* rearranges the sentence to be more direct and straightforward, so it is correct. Choice *A* needs a comma after *on*. Choice *B* introduces unnecessary commas. Choice *C* creates an incomplete sentence, since "Because I wasn't invested in what was going on" is a dependent clause.

14. C: Choice *C* is fluid and direct, making it the best revision. Choice *A* is incorrect because the construction is awkward and lacks parallel structure. Choice *B* is incorrect because of the unnecessary comma and period. Choice *D* is close, but its sequence is still awkward and overly complicated.

15. B: Choice *B* correctly adds a comma after *person* and cuts out the extraneous writing, making the sentence more streamlined. Choice *A* is poorly constructed, lacking proper grammar to connect the sections of the sentence correctly. Choice *C* inserts an unnecessary semicolon and doesn't enable this section to flow well with the rest of the sentence. Choice *D* is better but still unnecessarily long.

16. D: This sentence, though short, is a complete sentence. The only thing the sentence needs is an em dash after *easy*. In this sentence, the em dash works to add emphasis to the word *easy* and also acts in place of a colon, but in a less formal way. Therefore, Choice *D* is correct. Choices *A* and *B* lack the crucial comma, while Choice *C* unnecessarily breaks the sentence apart.

17. C: Choice *C* successfully fixes the construction of the sentence, changing *drawing* into *to draw*. Keeping the original sentence disrupts the flow, so Choice *A* is incorrect. Choice *B*'s use of *which* offsets the whole sentence. Choice *D* is incorrect because it unnecessarily expands the sentence content and makes it more confusing.

18. B: Choice *B* fixes the homophone issue. Because the author is talking about people, *their* must be used instead of *there*. This revision also appropriately uses the Oxford comma, separating and distinguishing *lives, world, and future*. Choice *A* uses the wrong homophone and is missing commas. Choice *C* neglects to fix these problems and unnecessarily changes the tense of *applies*. Choice *D* fixes the homophone but fails to properly separate *world* and *future*.

19. C: Choice *C* is correct because it fixes the core issue with this sentence: the singular *has* should not describe the plural *scientists*. Thus, Choice *A* is incorrect. Choices *B* and *D* add unnecessary commas.

20. D: Choice *D* correctly conveys the writer's intention of asking if, or *whether*, early perceptions of dinosaurs are still influencing people. Choice *A* makes no sense as worded. Choice *B* is better, but *how* doesn't make sense in context, because the author doesn't specifically explore how early perceptions influence understanding. Choice *C* adds unnecessary commas.

21. A: Choice *A* is correct, as the sentence does not require modification. Choices *B* and *C* implement extra punctuation unnecessarily, disrupting the flow of the sentence. Choice *D* incorrectly adds a comma in an awkward location.

22. B: Choice *B* is the strongest revision, as adding *to explore* is very effective in both shortening the sentence and maintaining, even enhancing, the point of the writer. To explore is to seek understanding in order to gain knowledge and insight, which coincides with the focus of the overall sentence. Choice *A* is not technically incorrect, but it is overcomplicated. Choice *C* is a decent revision, but the sentence could still be more condensed and sharpened. Choice *D* fails to make the sentence more concise and inserts unnecessary commas.

23. D: Choice *D* correctly applies a semicolon to introduce a new line of thought while remaining in a single sentence. The comma after *however* is also appropriately placed. Choice *A* is a run-on sentence. Choice *B* is also incorrect because the single comma is not enough to fix the sentence. Choice *C* adds commas around *uncertain* which are unnecessary.

24. B: Choice *B* not only fixes the homophone issue from *its*, which is possessive, to *it's*, which is a contraction of *it is*, but also streamlines the sentence by adding a comma and eliminating *and*. Choice *A* is incorrect because of these errors. Choices *C* and *D* only fix the homophone issue.

25. A: Choice *A* is correct, as the sentence is fine the way it is. Choices *B* and *C* add unnecessary commas, while Choice *D* uses the possessive *its* instead of the contraction *it's*.

26. C: Choice *C* is correct because the phrase *even likely* is flanked by commas, creating a kind of aside, which allows the reader to see this separate thought while acknowledging it as part of the overall sentence and subject at hand. Choice *A* is incorrect because it seems to ramble after *even* due to a missing comma after *likely*. Choice *B* is better but inserting a comma after *that* warps the flow of the writing. Choice *D* is incorrect because there must be a comma after *plausible*.

27. D: Choice *D* strengthens the overall sentence structure while condensing the words. This makes the subject of the sentence, and the emphasis of the writer, much clearer to the reader. Thus, while Choice *A* is technically correct, the language is choppy and over-complicated. Choice *B* is better but lacks the reference to a specific image of dinosaurs. Choice *C* introduces unnecessary commas.

28. B: Choice *B* correctly joins the two independent clauses. Choice *A* is decent, but "that would be" is too verbose for the sentence. Choice *C* incorrectly changes the semicolon to a comma. Choice *D* splits the clauses effectively but is not concise enough.

29. A: Choice *A* is correct, as the original sentence has no error. Choices *B* and *C* employ unnecessary semicolons and commas. Choice *D* would be an ideal revision, but it lacks the comma after *Ransom* that would enable the sentence structure to flow.

30. D: By reorganizing the sentence, the context becomes clearer with Choice *D*. Choice *A* has an awkward sentence structure and is less direct than Choice *D*. Choice *B* offers a revision that doesn't correspond well with the original sentence's intent. Choice *C* cuts out too much of the original content, losing the full meaning.

31. C: Choice *C* fixes the disagreement between the singular *this* and the plural *viewpoints*. Choice *A*, therefore, is incorrect. Choice *B* additionally introduces an unnecessary comma. In Choice *D*, *those* agrees with *viewpoints*, but neither agrees with *distinguishes*.

32. A: Choice *A* is direct and clear, without any punctuation errors. Choice *B* is well-written but too wordy. Choice *C* adds an unnecessary comma. Choice *D* is also well-written but much less concise than Choice *A*.

33. D: Choice *D* rearranges the sentence to improve clarity and impact, with *tempting* directly describing *idea*. On its own, Choice *A* has an unnecessary comma and a comma splice between two independent clauses. Choice *B* is better because it separates the clauses, but it keeps an unnecessary comma. Choice *C* is also an improvement but still has a comma splice.

34. B: Choice *B* is the best answer simply because the sentence makes it clear that Un-man takes over and possesses Weston. In Choice *A*, these events sounded like two different things, instead of an action and result. Choices *C* and *D* make this relationship clearer, but the revisions don't flow very well grammatically.

35. D: Changing the phrase *after this* to *then* makes the sentence less complicated and captures the writer's intent, making Choice *D* correct. Choice *A* is awkwardly constructed. Choices *B* and *C* misuse their commas and do not adequately improve the clarity.

36. B: By starting a new sentence, the run-on issue is eliminated, and a new line of reasoning can be seamlessly introduced, making Choice *B* correct. Choice *A* is thus incorrect. While Choice *C* fixes the run-on via a semicolon, a comma is still needed after *this*. Choice *D* contains a comma splice. The independent clauses must be separated by more than just a comma, even with the rearrangement of the second half of the sentence.

37. C: Choice *C* condenses the original sentence while being more active in communicating the emphasis on the changing times and media that the author is going for, so it is correct. Choice *A* is clunky because it lacks a comma after *today* to successfully transition into the second half of the sentence. Choice *B* inserts unnecessary commas. Choice *D* is a good revision of the underlined section, but not only does it not fully capture the original meaning, it also does not flow into the rest of the sentence.

38. B: Choice *B* clearly illustrates the author's point, with a well-placed semicolon that breaks the sentence into clearer, more readable sections. Choice *A* lacks punctuation. Choice *C* is incorrect because the period inserted after *question* forms an incomplete sentence. Choice *D* is a very good revision but does not make the author's point clearer than the original.

39. A: Choice *A* is correct: while the sentence seems long, it actually doesn't require any commas. The conjunction "that" successfully combines the two parts of the sentence without the need for additional punctuation. Choices *B* and *C* insert commas unnecessarily, incorrectly breaking up the flow of the sentence. Choice *D* alters the meaning of the original text by creating a new sentence, which is only a fragment.

40. C: Choice *C* correctly replaces *for* with *to*, the correct preposition for the selected area. Choice *A* is not the answer because of this incorrect preposition. Choice *B* is unnecessarily long and disrupts the original sentence structure. Choice *D* is also too wordy and lacks parallel structure.

41. D: Choice *D* is the answer because it inserts the correct punctuation to fix the sentence, linking the dependent and independent clauses. Choice *A* is therefore incorrect. Choice *B* is also incorrect since this revision only adds content to the sentence while lacking grammatical precision. Choice *C* overdoes the punctuation; only a comma is needed, not a semicolon.

42. B: Choice *B* correctly separates the section into two sentences and changes the word order to make the second part clearer. Choice *A* is incorrect because it is a run-on. Choice *C* adds an extraneous comma, while Choice *D* makes the run-on worse and does not coincide with the overall structure of the sentence.

43. C: Choice *C* is the best answer because of how the commas are used to flank *in earnest*. This distinguishes the side thought (*in earnest*) from the rest of the sentence. Choice *A* needs punctuation. Choice *B* inserts a semicolon in a spot that doesn't make sense, resulting in a fragmented sentence and lost meaning. Choice *D* is unnecessarily repetitive and creates a run-on.

44. A: Choice *A* is correct because the sentence contains no errors. The comma after *bias* successfully links the two halves of the sentence, and the use of *it's* is correct as a contraction of *it is*. Choice *B* creates a sentence fragment, while Choice *C* creates a run-on. Choice *D* incorrectly changes *it's* to the possessive *its*.

45. D: Choice *D* correctly inserts a comma after *however* and fixes *over use* to *overuse*—in this usage, it is one word. Choice *A* is therefore incorrect, as is Choice *B*. Choice *C* is a good revision but does not fit well with the rest of the sentence.

Math Test

1. B: To simplify this inequality, subtract 3 from both sides to get $-\frac{1}{2}x \geq -1$. Then, multiply both sides by -2 (remembering this flips the direction of the inequality) to get $x \leq 2$.

2. D: This problem involves a composition function, where one function is plugged into the other function. In this case, the $f(x)$ function is plugged into the $g(x)$ function for each x-value. The composition equation becomes:

$$g(f(x)) = 2^3 - 3(2)^2 - 2(2) + 6$$

Simplifying the equation gives the answer:

$$g(f(x)) = 8 - 3(4) - 2(2) + 6$$

$$8 - 12 - 4 + 6 = -2$$

3. B: A factor of 36 is any number that can be divided into 36 and have no remainder.

$$36 = 36 \times 1, 18 \times 2, 9 \times 4, \text{ and } 6 \times 6$$

Therefore, it has 7 unique factors: 36, 18, 9, 6, 4, 2, and 1.

4. D: A parabola of the form $y = \frac{1}{4f}x^2$ has a focus $(0, f)$.

Because $y = -9x^2$, set $-9 = \frac{1}{4f}$.

Solving this equation for f results in $f = -\frac{1}{36}$. Therefore, the coordinates of the focus are $\left(0, -\frac{1}{36}\right)$.

5. B: The volume of a cube is the length of the side cubed, and 3 inches cubed is 27 in³.

Choice *A* is not the correct answer because that is 2×3 inches.

Choice *C* is not the correct answer because that is 3×3 inches, and Choice *D* is not the correct answer because there was no operation performed.

6. B: The volume of a rectangular prism is the $length \times width \times height$, and $3\ cm \times 5\ cm \times 11\ cm$ is 165 cm³.

Choice *A* is not the correct answer because that is $3\ cm + 5\ cm + 11\ cm$. Choice *C* is not the correct answer because that is 15^2. Choice *D* is not the correct answer because that is $3cm \times 5cm \times 10cm$.

7. A: The volume of a cylinder is $\pi r^2 h$, and $\pi \times 5^2 \times 10$ is $250\ \pi\ in^3$.

Choice *B* is not the correct answer because that is $5^2 \times 2\pi$. Choice *C* is not the correct answer since that is $5in \times 10\pi$. Choice *D* is not the correct answer because that is $10^2 \times 2in$.

8. D: This system of equations involves one quadratic function and one linear function, as seen from the degree of each equation. One way to solve this is through substitution. Solving for y in the second equation yields $y = x + 2$.

Plugging this equation in for the y of the quadratic equation yields:

$$x^2 - 2x + x + 2 = 8$$

Simplifying the equation, it becomes $x^2 - x + 2 = 8$.

Setting this equal to zero and factoring, it becomes:

$$x^2 - x - 6 = 0 = (x - 3)(x + 2)$$

Solving these two factors for x gives the zeros $x = 3, -2$. To find the y-value for the point, each number can be plugged in to either original equation. Solving each one for y yields the points $(3, 5)$ and $(-2, 0)$.

9. B: The slope will be given by $\frac{1-0}{2-0} = \frac{1}{2}$.

The y-intercept will be 0, since it passes through the origin. Using slope-intercept form, the equation for this line is $y = \frac{1}{2}x$.

10. D: Recall the formula for area, area = length × width. The answer must be in square inches, so all values must be converted to inches. Half of a foot is equal to 6 inches. Therefore, the area of the rectangle is equal to:

$$6 \text{ in} \times \frac{11}{2} \text{ in} = \frac{66}{2} \text{ in}^2 = 33 \text{ in}^2$$

11. B: The table shows values that are increasing exponentially. The differences between the inputs are the same, while the differences in the outputs are changing by a factor of 2. The values in the table can be modeled by the equation $f(x) = 2^x$.

12. B: For the first card drawn, the probability of a King being pulled is $\frac{4}{52}$. Since this card isn't replaced, if a King is drawn first the probability of a King being drawn second is $\frac{3}{51}$. The probability of a King being drawn in both the first and second draw is the product of the two probabilities:

$$\frac{4}{52} \times \frac{3}{51} = \frac{12}{2652}$$

This fraction, when divided by 12, equals $\frac{1}{221}$.

13. C: The picture demonstrates Angle-Side-Angle congruence. Choice A and B are incorrect because the picture does not show Side-Side-Side congruence and angles alone cannot prove congruence. Choice D is not the correct answer because there is already enough information to prove congruence.

14. D: The expression is three times the sum of twice a number and 1, which is $3(2x + 1)$. Then, 6 is subtracted from this expression.

15. C: To solve this equation, square both sides to eliminate the radical, resulting in $x + 5 = 25$. Subtracting 5 from both sides to solve for x gives $x = 20$.

16. A: To expand a squared binomial, it's necessary to use the FOIL (First, Outer, Inner, Last) method:

$$(2x - 4y)^2$$

$$2x \times 2x + 2x(-4y) + (-4y)(2x) + (-4y)(-4y)$$

$$4x^2 - 8xy - 8xy + 16y^2$$

$$4x^2 - 16xy + 16y^2$$

17. B: The zeros of this function can be found by using the quadratic formula:

$$x = \frac{-b \pm \sqrt{b^2 - 4ac}}{2a}$$

Identifying a, b, and c can also be done from the equation because it is in standard form. The formula becomes:

$$x = \frac{0 \pm \sqrt{0^2 - 4(1)(4)}}{2(1)} = \frac{\sqrt{-16}}{2}$$

Since there is a negative underneath the radical, the answer is a complex number:

$$x = \pm 2i$$

18. D: The expression is simplified by collecting like terms. Terms with the same variable and exponent are like terms, and their coefficients can be added.

19. A: To simplify this expression, first line up the fractions:

$$\frac{15}{23} \times \frac{54}{127}$$

Multiply across the top and across the bottom to find the numerator and denominator:

$$\frac{15 \times 54}{23 \times 127} = \frac{810}{2921}$$

Because the numerator and denominator do not share any common factors, the resulting fraction cannot be reduced.

20. A: Finding the product means distributing one polynomial to the other so that each term in the first is multiplied by each term in the second. Then, like terms can be collected. Multiplying the factors yields the expression:

$$20x^3 + 4x^2 + 24x - 40x^2 - 8x - 48$$

Collecting like terms means adding the x^2 terms and adding the x terms. The final answer after simplifying the expression is:

$$20x^3 - 36x^2 + 16x - 48$$

21. B: The equation can be solved by factoring the numerator into $(x + 6)(x - 5)$. Since that same factor $(x - 5)$ exists on top and bottom, that factor cancels. This leaves the equation $x + 6 = 11$. Solving the equation gives the answer $x = 5$. When this value is plugged into the equation, it yields a zero in the denominator of the fraction. Since this is undefined, there is no solution.

22. A: The common denominator here will be 4x. Rewrite these fractions as:

$$\frac{3}{x} + \frac{5u}{2x} - \frac{u}{4}$$

$$\frac{12}{4x} + \frac{10u}{4x} - \frac{ux}{4x}$$

$$\frac{12 + 10u - ux}{4x}$$

23. B: There are two zeros for the given function. They are $x = 0, -2$. The zeros can be found a number of ways, but this particular equation can be factored into:

$$f(x) = x(x^2 + 4x + 4) = x(x + 2)(x + 2)$$

By setting each factor equal to zero and solving for x, there are two solutions. On a graph, these zeros can be seen where the line crosses the x-axis.

24. A: The equation is *even* because $f(-x) = f(x)$. Plugging in a negative value will result in the same answer as when plugging in the positive of that same value. The function:

$$f(-2) = \frac{1}{2}(-2)^4 + 2(-2)^2 - 6$$

$$8 + 8 - 6 = 10$$

This function yields the same value as:

$$f(2) = \frac{1}{2}(2)^4 + 2(2)^2 - 6$$

$$8 + 8 - 6 = 10$$

25. B: The perimeter of a rectangle is the sum of all four sides. Therefore, the answer is:

$$P = 14 + 8\frac{1}{2} + 14 + 8\frac{1}{2}$$

$$14 + 14 + 8 + \frac{1}{2} + 8 + \frac{1}{2}$$

45 square inches

26. B: $12 \times 750 = 9{,}000$. Therefore, there are 9,000 milliliters of water, which must be converted to liters. 1,000 milliliters equals 1 liter; therefore, 9 liters of water are purchased.

27. B: Because this isn't a right triangle, SOHCAHTOA can't be used. However, the law of cosines can be used. Therefore:

$$c^2 = a^2 + b^2 - 2ab \cos C$$

$$19^2 + 26^2 - 2 \times 19 \times 26 \times \cos 42°$$

$$302.773$$

Taking the square root and rounding to the nearest tenth results in $c = 17.4$.

28. C: Because the triangles are similar, the lengths of the corresponding sides are proportional. Therefore:

$$\frac{30 + x}{30} = \frac{22}{14} = \frac{y + 15}{y}$$

This results in the equation $14(30 + x) = 22 \times 30$ which, when solved, gives $x = 17.1$. The proportion also results in the equation $14(y + 15) = 22y$ which, when solved, gives $y = 26.3$.

29. B: The technique of completing the square must be used to change:

$$4x^2 + 4y^2 - 16x - 24y + 51 = 0$$

into the standard equation of a circle. First, the constant must be moved to the right-hand side of the equals sign, and each term must be divided by the coefficient of the x^2 term (which is 4). The x and y terms must be grouped together to obtain:

$$x^2 - 4x + y^2 - 6y = -\frac{51}{4}$$

Then, the process of completing the square must be completed for each variable. This gives:

$$(x^2 - 4x + 4) + (y^2 - 6y + 9)$$

$$-\frac{51}{4} + 4 + 9$$

The equation can be written as:

$$(x - 2)^2 + (y - 3)^2 = \frac{1}{4}$$

Therefore, the center of the circle is (2, 3) and the radius is:

$$\sqrt{\frac{1}{4}} = \frac{1}{2}$$

30. A: Operations within the parentheses must be completed first. Then, division is completed. Finally, addition is the last operation to complete. When adding decimals, digits within each place value are added together. Therefore, the expression is evaluated as:

$$(2 \times 20) \div (7 + 1) + (6 \times 0.01) + (4 \times 0.001)$$

$$40 \div 8 + 0.06 + 0.004 = 5 + 0.06 + 0.004 = 5.064$$

31. C: A dollar contains 20 nickels. Therefore, if there are 12 dollars' worth of nickels, there are $12 \times 20 = 240$ nickels. Each nickel weighs 5 grams. Therefore, the weight of the nickels is $240 \times 5 = 1,200$ grams. Adding in the weight of the empty piggy bank, the filled bank weighs 2,250 grams.

32. D: To find Denver's total snowfall, 3 must be multiplied times $27\frac{3}{4}$. In order to easily do this, the mixed number should be converted into an improper fraction.

$$27\frac{3}{4} = \frac{27 \times 4 + 3}{4} = \frac{111}{4}$$

Therefore, Denver had approximately $\frac{3 \times 111}{4} = \frac{333}{4}$ inches of snow. The improper fraction can be converted back into a mixed number through division.

$$\frac{333}{4} = 83\frac{1}{4} \text{ inches}$$

33. D: $x \leq -5$. When solving a linear equation or inequality:

Distribution is performed if necessary: $-3(x + 4)$, or $-3x - 12 \geq x + 8$. This means that any like terms on the same side of the equation/inequality are combined.

The equation/inequality is manipulated to get the variable on one side. In this case, subtracting x from both sides produces $-4x - 12 \geq 8$.

The variable is isolated using inverse operations to undo addition/subtraction. Adding 12 to both sides produces $-4x \geq 20$.

The variable is isolated using inverse operations to undo multiplication/division. Remember if dividing by a negative number, the relationship of the inequality reverses, so the sign is flipped. In this case, dividing by -4 on both sides produces $x \leq -5$.

34. C: $y = 40x + 300$. In this scenario, the variables are the number of sales and Karen's weekly pay. The weekly pay depends on the number of sales. Therefore, weekly pay is the dependent variable (y), and the number of sales is the independent variable (x). Each pair of values from the table can be written as an ordered pair (x, y): $(2, 380), (7, 580), (4, 460), (8, 620)$. The ordered pairs can be substituted into the equations to see which creates true statements (both sides equal) for each pair. Even if one ordered pair produces equal values for a given equation, the other three ordered pairs must be checked. The only equation which is true for all four ordered pairs is $y = 40x + 300$:

$$380 = 40(2) + 300 \rightarrow 380 = 380$$

$$580 = 40(7) + 300 \rightarrow 580 = 580$$

$$460 = 40(4) + 300 \rightarrow 460 = 460$$

$$620 = 40(8) + 300 \rightarrow 620 = 620$$

35. C: The area of the shaded region is the area of the square, minus the area of the circle. The area of the circle will be πr^2. The side of the square will be $2r$, so the area of the square will be $4r^2$. Therefore, the difference is:

$$4r^2 - \pi r^2 = (4 - \pi)r^2$$

36. B: The car is traveling at a speed of five meters per second. On the interval from one to three seconds, the position changes by ten meters. By making this change in position over time into a rate, the speed becomes ten meters in two seconds or five meters in one second.

37. B: For an ordered pair to be a solution to a system of inequalities, it must make a true statement for BOTH inequalities when substituting its values for x and y. Substituting (-3,-2) into the inequalities produces $(-2) > 2(-3) - 3$, which is $-2 > -9$, and $(-2) < -4(-3) + 8$, or $-2 < 20$. Both are true statements.

38. D: The shape of the scatterplot is a parabola (U-shaped). This eliminates Choices *A* (a linear equation that produces a straight line) and *C* (an exponential equation that produces a smooth curve upward or downward). The value of a for a quadratic function in standard form ($y = ax^2 + bx + c$) indicates whether the parabola opens up (U-shaped) or opens down (upside-down U). A negative value for a produces a parabola that opens down; therefore, Choice *B* can also be eliminated.

39. B: According to the order of operations, multiplication and division must be completed first from left to right. Then, addition and subtraction are completed from left to right. Therefore:

$$9 \times 9 \div 9 + 9 - 9 \div 9$$

$$81 \div 9 + 9 - 9 \div 9$$

$$9 + 9 - 9 \div 9$$

$$9 + 9 - 1$$

$$18 - 1$$

$$17$$

40. B: First, subtract 9 from both sides to isolate the radical. Then, cube each side of the equation to obtain:

$$2x + 11 = 27$$

Subtract 11 from both sides, and then divide by 2. The result is $x = 8$. Plug 8 back into the original equation to obtain the true statement to check the answer:

$$\sqrt[3]{16 + 11} + 9 = 12$$

$$\sqrt[3]{27} + 9 = 12$$

$$3 + 9 = 12$$

41. A: 13 nurses. Using the given information of 1 nurse to 25 patients and 325 patients, set up an equation to solve for number of nurses (N):

$$\frac{N}{325} = \frac{1}{25}$$

Multiply both sides by 325 to get N by itself on one side:

$$\frac{N}{1} = \frac{325}{25} = 13 \; nurses$$

42. D: 290 beds. Using the given information of 2 beds to 1 room and 145 rooms, set up an equation to solve for number of beds (B):

$$\frac{B}{145} = \frac{2}{1}$$

Multiply both sides by 145 to get B by itself on one side.

$$\frac{B}{1} = \frac{290}{1} = 290 \; beds$$

43. C: $x = 150$. Set up the initial equation:

$$\frac{2x}{5} - 1 = 59$$

Add 1 to both sides:

$$\frac{2x}{5} - 1 + 1 = 59 + 1$$

Multiply both sides by $\frac{5}{2}$:

$$\frac{2x}{5} \times \frac{5}{2} = 60 \times \frac{5}{2} = 150$$

$$x = 150$$

44. C: $51.93. First, list the givens:

$$Tax = 6.0\% = 0.06$$

$$Sale = 50\% = 0.5$$

$$Hat = \$32.99$$

$$Jersey = \$64.99$$

Calculate the sale prices for hats and jerseys:

$$Hat \; sale = 0.5 \, (32.99) = 16.495$$

$$Jersey \; sale = 0.5 \, (64.99) = 32.495$$

Total the sale prices:

$$Hat\ sale + jersey\ sale = 16.495 + 32.495 = 48.99$$

Finally, calculate the sales tax and add it to the total sale prices:

$$Total\ after\ tax = 48.99 + (48.99 \times 0.06) = \$51.93$$

45. D: $0.45. First, list the givens:

$$Store\ coffee = \$1.23/lb$$

$$Local\ roaster\ coffee = \$1.98/1.5\ lb$$

Calculate the cost for 5 pounds of store brand coffee.

$$\frac{\$1.23}{1\ lb} \times 5\ lb = \$6.15$$

Calculate the cost for 5 pounds of the local roaster's coffee:

$$\frac{\$1.98}{1.5\ lb} \times 5\ lb = \$6.60$$

Subtract to find the difference in price for 5 pounds of coffee:

$$\$6.60 - \$6.15 = \$0.45$$

46. D: $3,325. First, list the givens:

$$1,800\ ft = \$2,000$$

$$Cost\ after\ 1,800\ ft = \$1.00/ft$$

Find how many feet left after the first 1,800 ft:

$$3125\ ft - 1,800\ ft = 1325\ ft$$

Calculate the cost for the feet over 1,800 ft:

$$1,325\ ft \times \frac{\$1.00}{1\ ft} = \$1,325$$

Total for entire cost:

$$\$2,000 + \$1,325 = \$3,325$$

47. A: 12. Calculate how many gallons the bucket holds:

$$11.4\ L \times \frac{1\ gal}{3.8\ L} = 3\ gal$$

Now how many buckets to fill the pool which needs 35 gallons:

$$\frac{35 \ gal}{3 \ gal} = 11.67$$

Since the amount is more than 11 but less than 12, we must fill the bucket 12 times.

48. D: Three girls for every two boys can be expressed as a ratio: 3:2. This can be visualized as splitting the school into 5 groups: 3 girl groups and 2 boy groups. The number of students in each group can be found by dividing the total number of students by 5:

$$\frac{650 \ students}{5 \ groups} = \frac{130 \ students}{group}$$

To find the total number of girls, multiply the number of students per group (130) by the number of girl groups in the school (3). This equals 390, Choice *D*.

49. C: The volume of a pyramid is *length* × *width* × *height*, divided by 3, and 6 × 6 × 9, divided by 3 is 108 in³. Choice *A* is incorrect because 324 in³ is *length* × *width* × *height* without dividing by 3. Choice *B* is incorrect because 6 is used for height instead of 9 (6 × 6 × 6) and divided by 3 to get 72 in³. Choice *D* is incorrect because 18 in³ is 6 × 9, divided by 3, and leaving out a 6.

50. A: 22%. Converting from a fraction to a percentage generally involves two steps. First, the fraction needs to be converted to a decimal:

$$\frac{2}{9} = 0.\overline{22}$$

The top line indicates that the decimal actually goes on forever with an endless amount of 2's.

Second, the decimal needs to be moved two places to the right to convert to a percent:

$$22\%$$

51. A: The volume of a cone is $\pi r^2 h$, divided by 3, and $\pi \times 10^2 \times 12$, divided by 3, is 400 cm³. Choice *B* is $10^2 \times 2$. Choice *C* is incorrect because it is 10 × 12. Choice *D* is also incorrect because that is $10^2 + 40$.

52. A: 37.5%. Solve this by setting up the percent formula:

$$\frac{3}{8} = \frac{\%}{100}$$

Multiply 3 by 100 to get 300. Then divide 300 by 8:

$$\frac{300}{8} = 37.5\%$$

Note that with the percent formula, 37.5 is automatically a percentage and does not need to have any further conversions.

53. C: 216 cm. Because area is a two-dimensional measurement, the dimensions are multiplied by a scale that is squared to determine the scale of the corresponding areas. The dimensions of the rectangle are multiplied by a scale of 3. Therefore, the area is multiplied by a scale of 3^2 (which is equal to 9):

$$24 \ cm \times 9 = 216 \ cm$$

54. Add 3 to both sides to get $4x = 8$. Then, divide both sides by 4 to get $x = 2$.

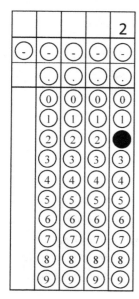

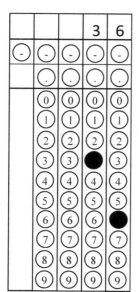

55. To solve this correctly, keep in mind the order of operations with the mnemonic PEMDAS (Please Excuse My Dear Aunt Sally). This stands for Parentheses, Exponents, Multiplication, Division, Addition, Subtraction. Taking it step by step, solve inside the parentheses first:

$$4 \times 7 + (4)^2 \div 2$$

Then, apply the exponent:

$$4 \times 7 + 16 \div 2$$

Multiplication and division are both performed next:

$$28 + 8 = 36$$

Addition and subtraction are done last. The solution is 36.

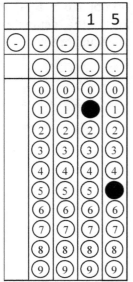

56. Follow the order of operations in order to solve this problem. Simplify the radicals within the parentheses first, and then follow the remainder as usual:

$$(6 \times 4) - 9$$

This equals $24 - 9$, or 15.

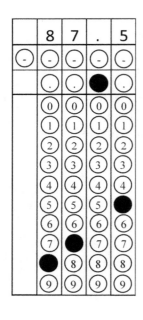

57. For an even number of total values, the *median* is calculated by finding the *mean* or average of the two middle values once all values have been arranged in ascending order from least to greatest. In this case, $(92 + 83) \div 2$ would equal the median 87.5.

58. The formula for the perimeter of a rectangle is $P = 2L + 2W$, where P is the perimeter, L is the length, and W is the width. The first step is to substitute all of the data into the formula:

$$36 = 2(12) + 2W$$

Simplify by multiplying 2×12:

$$36 = 24 + 2W$$

Simplifying this further by subtracting 24 on each side, which gives:

$$36 - 24 = 24 - 24 + 2W$$
$$12 = 2W$$

Divide by 2:

$$6 = W$$

The width is 6 cm. Remember to test this answer by substituting this value into the original formula:

$$36 = 2(12) + 2(6)$$

SAT Practice Test #2

Reading Test

Fiction

Questions 1–10 are based on the following passage:

A lane was forthwith opened through the crowd of spectators. Preceded by the beadle, and attended by an irregular procession of stern-browed men and unkindly visaged women, Hester Prynne set forth towards the place appointed for her punishment. A crowd of eager and curious school-boys, understanding little of the matter in hand, except that it gave them a half-holiday, ran before her progress, turning their heads continually to stare into her face, and at the winking baby in her arms, and at the ignominious letter on her breast. It was no great distance, in those days, from the prison-door to the market-place. Measured by the prisoner's experience, however, it might be reckoned a journey of some length; for, haughty as her demeanor was, she perchance underwent an agony from every footstep of those that thronged to see her, as if her heart had been flung into the street for them all to spurn and trample upon. In our nature, however, there is a provision, alike marvelous and merciful, that the sufferer should never know the intensity of what he endures by its present torture, but chiefly by the pang that rankles after it. With almost a serene deportment, therefore, Hester Prynne passed through this portion of her ordeal, and came to a sort of scaffold, at the western extremity of the market-place. It stood nearly beneath the eaves of Boston's earliest church, and appeared to be a fixture there.

In fact, this scaffold constituted a portion of a penal machine, which now, for two or three generations past, has been merely historical and traditionary among us, but was held, in the old time, to be as effectual an agent, in the promotion of good citizenship, as ever was the guillotine among the terrorists of France. It was, in short, the platform of the pillory; and above it rose the framework of that instrument of discipline, so fashioned as to confine the human head in its tight grasp, and thus hold it up to the public gaze. The very ideal of ignominy was embodied and made manifest in this contrivance of wood and iron. There can be no outrage, methinks, against our common nature,—whatever be the delinquencies of the individual,—no outrage more flagrant than to forbid the culprit to hide his face for shame; as it was the essence of this punishment to do. In Hester Prynne's instance, however, as not unfrequently in other cases, her sentence bore, that she should stand a certain time upon the platform, but without undergoing that gripe about the neck and confinement of the head, the proneness to which was the most devilish characteristic of this ugly engine. Knowing well her part, she ascended a flight of wooden steps, and was thus displayed to the surrounding multitude, at about the height of a man's shoulders above the street.

The scene was not without a mixture of awe, such as must always invest the spectacle of guilt and shame in a fellow-creature, before society shall have grown corrupt enough to smile, instead of shuddering, at it. The witnesses of Hester Prynne's disgrace had not yet passed beyond their simplicity. They were stern enough to look upon her death, had that been the sentence, without a murmur at its severity, but had none of the heartlessness of another social state, which would find only a theme for jest in an exhibition like the present. Even had there been a disposition to turn the matter into ridicule, it must have been repressed and overpowered by the solemn presence of men no less dignified than the Governor, and several of his counsellors, a judge, a general, and the ministers of the town; all of whom sat or stood in a balcony of the meeting-house, looking down upon the platform. When such personages could constitute a part of the spectacle, without risking the majesty or reverence of rank and office, it was safely to be inferred that the infliction of a legal sentence would have an earnest and effectual meaning. Accordingly, the crowd was sombre and grave. The unhappy culprit sustained herself as best a woman might, under the heavy weight of a thousand unrelenting eyes, all fastened upon her, and concentrated at her bosom. It was almost intolerable to be borne. Of an impulsive and passionate nature, she had fortified herself to encounter the stings and venomous stabs of public contumely, wreaking itself in every variety of insult; but there was a quality so much more terrible in the solemn mood of the popular mind, that she longed rather to behold all those rigid countenances contorted with scornful merriment, and herself the object. Had a roar of laughter burst from the multitude,—each man, each woman, each little shrill-voiced child, contributing their individual parts,—Hester Prynne might have repaid them all with a bitter and disdainful smile. But, under the leaden infliction which it was her doom to endure, she felt, at moments, as if she must needs shriek out with the full power of her lungs, and cast herself from the scaffold down upon the ground, or else go mad at once.

Excerpt from *The Scarlet Letter*, Nathaniel Hawthorne, 1878

1. Based on the first paragraph, what might Hester Prynne feel on her walk from prison to marketplace?
    a. Anger
    b. Fear
    c. Agony
    d. Proud

2. Based on the passage, what is the spectators' mood?
    a. Grave
    b. Amused
    c. Vengeful
    d. Outraged

3. The passage includes a description of the stockade but states it was not part of Hester's punishment. What can we infer from its inclusion?
    a. They didn't use it because she was a woman.
    b. The townspeople would have preferred its use.
    c. The threat of it was often a deterrent.
    d. The punishment could have been harsher.

4. Based on paragraph 3 of the passage, why was Prynne not mocked or ridiculed?
    a. Others had also committed crimes.
    b. She appeared too proud.
    c. Dignitaries were present.
    d. The crowd wanted to see her hanged.

5. In the final paragraph, what does the word "disdainful" mean?
   a. Admiring
   b. Contemptuous
   c. Sympathetic
   d. Mournful

6. In the first paragraph, why might a prisoner find the walk from prison to marketplace longer than the actual distance?
   a. The spectators watching cause the prisoner an emotional burden.
   b. The prisoner's ankles were shackled.
   c. Prisoners are forced to walk slowly.
   d. Kids keep running in front, taunting.

7. How did the crowd's reaction defy Prynne's expectations?
   a. She expected silence, but they cheered.
   b. She expected cheering, but they mocked her.
   c. She expected physical attacks, but they laughed.
   d. She expected verbal attacks but got none.

8. What can we conclude about Prynne's need to shriek and throw herself down or go mad?
   a. The weight of the punishment has finally hit her.
   b. Someone has attacked her.
   c. She has finally been placed in the stockade.
   d. She wants to respond to the crowd.

9. Over the course of the passage, the main point of view shifts among…
   a. A witness, the judge, and Hester Prynne
   b. Hester Prynne's experience, the crowd's experience, and Prynne's husband
   c. The narrator's viewpoint, Hester Prynne, and the crowd
   d. Hester Prynne, members of the crowd, and the town historian

10. Based on the passage, Hester's character can best be described as:
    a. Tenacious
    b. Ashamed
    c. Bashful
    d. Enraged

History/Social Studies

Questions 11–20 are based on the following passage:

The basic problem confronting the world today, as I said in the beginning, is the preservation of human freedom for the individual and consequently for the society of which he is a part. We are fighting this battle again today as it was fought at the time of the French Revolution and as the time of the American Revolution. The issue of human liberty is as decisive now as it was then. I want to give you my conception of what is meant in my country by freedom of the individual.

Long ago in London during a discussion with Mr. Vyshinsky, he told me there was no such thing as freedom for the individual in the world. All freedom of the individual was conditioned by the rights of other individuals. That of course, I granted. I said: "We approach the question from a different point of view: we here in the United Nations are trying to develop ideals which will be broader in outlook, which will consider first the rights of man, which will consider what makes man more free; not governments, but man."

The totalitarian state typically places the will of the people second to decrees promulgated by a few men at the top.

Naturally there must always be consideration of the rights of others; but in a democracy this is not a restriction. Indeed, in our democracies we make our freedoms secure because each of us is expected to respect the rights of others and we are free to make our own laws. Freedom for our peoples is not only a right, but also a tool. Freedom of speech, freedom of the press, freedom of information, freedom of assembly—these are not just abstract ideals to us; they are tools with which we create a way of life, a way of life in which we can enjoy freedom.

Sometimes the processes of democracy are slow, and I have known some of our leaders to say that a benevolent dictatorship would accomplish the ends desired in a much shorter time than it takes to go through the democratic processes of discussion and the slow formation of public opinion. But there is no way of insuring that a dictatorship will remain benevolent or that power once in the hands of a few will be returned to the people without struggle or revolution. This we have learned by experience and we accept the slow processes of democracy because we know that shortcuts compromise principles on which no compromise is possible.

The final expression of the opinion of the people with us is through free and honest elections, with valid choices on basic issues and candidates. The secret ballot is an essential to free elections but you must have a choice before you. I have heard my husband say many times that a people need never lose their freedom if they kept their right to a secret ballot and if they used that secret ballot to the full. Basic decisions of our society are made through the expressed will of the people. That is why when we see these liberties threatened, instead of falling apart, our nation becomes unified and our democracies come together as a unified group in spite of our varied backgrounds and many racial strains.

In the United States we have a capitalistic economy. That is because public opinion favors that type of economy under the conditions in which we live. But we have imposed certain restraints; for instance, we have antitrust laws. These are the legal evidence of the determination of the American people to maintain an economy of free competition and not to allow monopolies to take away the people's freedom.

Excerpt from Eleanor Roosevelt's "The Struggle for Human Rights," September 28, 1948

11. What does the writer say about freedom as a tool?
    a. Dictators can use it to manipulate populations.
    b. It's not just a concept; it can help achieve a desired lifestyle.
    c. It can be abused and misused by people.
    d. Only skilled individuals should be able to wield such a powerful tool.

12. According to Roosevelt, why do some people argue in favor of dictatorship?
    a. Dictatorships need not be tyrannical.
    b. Dictators come to power for a reason.
    c. Some dictators have been quite successful and popular.
    d. They believe that a dictator could achieve goals could more quickly.

13. According to the passage, what is the goal of antitrust laws?
    a. To establish trust between governments and populations
    b. To force businesses to be fair to consumers
    c. To foster economic competition
    d. To allow businesses to compete with the government

14. In the final sentence of the second paragraph, what does the word "promulgated" mean?
    a. Stifled
    b. Whispered
    c. Promoted
    d. Attacked

15. What is the main idea of this passage?
    a. We need democracy to maintain and protect individual freedom.
    b. Dictatorships can be benevolent and should be considered viable.
    c. Elections guarantee democracy and require secret ballots.
    d. Capitalism directly supports democracy and so should be encouraged.

16. Why does Roosevelt refer to both the French and American Revolutions?
    a. To demonstrate her understanding of history
    b. To draw a comparison to successful fights for freedom and democracy
    c. To highlight the fact that she is speaking to both American and French citizens
    d. To emphasize that she is speaking on the anniversary of both events

17. The tone of the passage is:
    a. Earnest
    b. Satirical
    c. Accusatory
    d. Conciliatory

18. Based on the passage, we can conclude that Roosevelt values:
    a. The relationship between France and America
    b. Capitalistic economy
    c. Knowledge of history
    d. Freedom and democracy

19. What is Roosevelt's concern about a "benevolent dictator"?
    a. The US has never had a dictator, so she is unfamiliar with them.
    b. The world is still recovering from World War II.
    c. There is no guarantee that a dictator will remain benevolent.
    d. There is no way a dictator will be benevolent.

20. Why does Roosevelt assert that a "secret" ballot is important?
    a. Public votes were too raucous an event.
    b. It decreases the chance for bribery.
    c. Privacy secures freedom of choice.
    d. It ensures physical safety.

History/Social Studies

Questions 21–30 are based on the following passage:

At times history and fate meet at a single time in a single place to shape a turning point in man's unending search for freedom. So it was at Lexington and Concord. So it was a century ago at Appomattox. So it was last week in Selma, Alabama. There, long-suffering men and women peacefully protested the denial of their rights as Americans. Many were brutally assaulted. One good man, a man of God, was killed.

There is no cause for pride in what has happened in Selma. There is no cause for self-satisfaction in the long denial of equal rights of millions of Americans. But there is cause for hope and for faith in our democracy in what is happening here tonight. For the cries of pain and the hymns and protests of oppressed people have summoned into convocation all the majesty of this great government—the government of the greatest nation on earth. Our mission is at once the oldest and the most basic of this country: to right wrong, to do justice, to serve man.

In our time we have come to live with the moments of great crisis. Our lives have been marked with debate about great issues—issues of war and peace, issues of prosperity and depression. But rarely in any time does an issue lay bare the secret heart of America itself. Rarely are we met with a challenge, not to our growth or abundance, or our welfare or our security, but rather to the values, and the purposes, and the meaning of our beloved nation.

The issue of equal rights for American Negroes is such an issue.

And should we defeat every enemy, and should we double our wealth and conquer the stars, and still be unequal to this issue, then we will have failed as a people and as a nation. For with a country as with a person, "What is a man profited, if he shall gain the whole world, and lose his own soul?"

There is no Negro problem. There is no Southern problem. There is no Northern problem. There is only an American problem. And we are met here tonight as Americans—not as Democrats or Republicans. We are met here as Americans to solve that problem.

This was the first nation in the history of the world to be founded with a purpose. The great phrases of that purpose still sound in every American heart, North and South: "All men are created equal," "government by consent of the governed," "give me liberty or give me death." Well, those are not just clever words, or those are not just empty theories. In their name Americans have fought and died for two centuries, and tonight around the world they stand there as guardians of our liberty, risking their lives.

Those words are a promise to every citizen that he shall share in the dignity of man. This dignity cannot be found in a man's possessions; it cannot be found in his power, or in his position. It really rests on his right to be treated as a man equal in opportunity to all others. It says that he shall share in freedom, he shall choose his leaders, educate his children, provide for his family according to his ability and his merits as a human being. To apply any other test—to deny a man his hopes because of his color, or race, or his religion, or the place of his birth is not only to do injustice, it is to deny America and to dishonor the dead who gave their lives for American freedom.

Except from Lyndon Johnson's "Address to Joint Session of Congress," March 15, 1965

21. What was being protested in Selma?
    a. The closing of a bridge
    b. Denial of rights
    c. A murder
    d. Loss of wealth

22. Based on this passage, why does Johnson suggest there is hope in the events of Selma?
    a. Police responded immediately to the situation.
    b. Lots of citizens showed up to protest.
    c. It forced attention on the issue and sparked conversations about it.
    d. Despite the violence, only one life was lost.

23. How does Johnson suggest this issue is different from other debates?
    a. Other debates haven't led to protests.
    b. It is easier to solve.
    c. It is more difficult to solve.
    d. It concerns our values.

24. Based on Johnson's words, where does dignity lie?
    a. In equality
    b. In freedom to protest
    c. In American might
    d. In American abundance

25. What best describes the tone of Johnson's speech?
    a. Righteous
    b. Quixotic
    c. Scathing
    d. Vindictive

26. What does Johnson suggest will indicate failure, even among amazing successes?
    a. Failure to reach the moon
    b. A loss of wealth
    c. Military defeat
    d. Moral or spiritual decay

27. In the second paragraph of this passage, what does the word "convocation" mean?
    a. A vocal duet
    b. A second career
    c. A gathering or assembly
    d. Closure

28. Based on the examples Johnson provides in the next to last paragraph of the passage, what can we conclude about the purpose for which America was founded?
    a. To fight wars and protect nations
    b. To lead the world
    c. To foster equality, democracy, and freedom
    d. To establish a country that others envy

29. Based on context, who are the "guardians of liberty" Johnson refers to?
    a. American citizens
    b. American troops
    c. American government
    d. Selma protestors

30. Which of the following best encapsulates the purpose of this speech?
    a. To entertain
    b. To describe
    c. To teach
    d. To persuade

31. What does the phrase "government by consent of the governed" mean?

    a. The governed people must like their government.
    b. The governed people must choose their government.
    c. The government must treat the governed people fairly.
    d. The government is not required to hold elections.

Science

Questions 32–42 are based on the following passage:

These community assessments conducted during the Zika outbreak, hurricane responses, and hurricane recovery in U.S. Virgin Islands (USVI) found that households were more concerned about contracting mosquito-borne diseases shortly after the Zika outbreak than during the hurricane response and hurricane recovery, even though reported mosquito biting activity increased, and environmental conditions were more favorable for mosquito breeding and exposure to bites following the hurricanes. In addition, although mosquito-borne diseases are endemic in USVI, and the population might be aware of the risk, households had concerns after the hurricanes that did not exist during the Zika outbreak, such as lack of shelter, clean water, and electricity. These differing levels of concern did not, however, change the community's support for mosquito spraying, although support for specific spray methods varied.

VIDOH used the Community Assessments for Public Health Emergency Response (CASPERs) data to make real-time outbreak and hurricane response decisions to improve mosquito bite prevention, mosquito control, and community education. For example, because the percentage of households concerned about contracting mosquito-borne diseases declined after the hurricanes compared with during the Zika outbreak response, VIDOH hurricane response education campaigns prioritized household-level mosquito bite prevention. The differing levels of support for various spray methods were also recognized and considered during decision making. For example, these data, along with unique environmental considerations, were used by the administration in place during the responses and recovery to determine backpack spraying to be the only acceptable option.

The CASPER is a useful tool for assessing mosquito-borne disease risk factors and creating immediately useable data to guide vector-related public health campaigns. According to CDC's internal CASPER database, a limited number of CASPERs have been conducted that assess mosquito bite prevention- and control-related factors, such as knowledge of mosquito-borne diseases; ways to protect against mosquito bites; and how to identify, quantify, and manage potential mosquito breeding sites. Even fewer CASPERs have focused solely on mosquitoes. A CASPER in Long Beach, California, during a Zika outbreak identified the need for increased mosquito abatement. In two areas of Texas, CASPERs successfully assessed the prevalence of vector-borne disease risk factors and the communities' knowledge of mosquito bite prevention and Zika virus. A CASPER conducted in American Samoa identified increased vector problems and the need for vector control after a tsunami.

Not only is CASPER an important tool for emergency response and recovery, it is also useful for collecting community public health information unrelated to an emergency. Vector control programs can use CASPERs during nonemergency situations to enhance and increase operation efficacy by evaluating the effectiveness of community campaigns and understanding community knowledge, attitudes, and practices.

Seger KR, Roth J Jr., Schnall AH, Ellis BR, Ellis EM. Community Assessments for Mosquito Prevention and Control Experiences, Attitudes, and Practices — U.S. Virgin Islands, 2017 and 2018. MMWR Morb Mortal Wkly Rep 2019;68:500–504. DOI: http://dx.doi.org/10.15585/mmwr.mm6822a3external icon.

32. Based on the passage, how did residents' self-assessment of their risk differ from the actual risk?
    a. Residents rated risk higher during the Zika outbreak than after the hurricane.
    b. Residents saw no risk at all from Zika.
    c. Residents saw no risk after the hurricane.
    d. Residents believed the risks to be equivalent.

33. The research in the first paragraph attempts to draw a correlation between which two things?
    a. Zika and hurricanes
    b. Resident concerns and Zika spread
    c. Hurricanes and mosquito spraying
    d. Basic necessities and mosquito-borne diseases

34. Which factor of Zika response have CASPERs in the U.S. Virgin Islands assessed?
    a. Hurricane response
    b. Swamp size
    c. Bite prevention
    d. Backpack spraying

35. Based on the passage, what does the term "vector-related" mean?
    a. Related to a region of a country
    b. Concerning avenues or pathways through which disease is spread
    c. Concerning a treatment area
    d. Related to lines in space

36. Based on context, what does the word "abatement" mean?
    a. Increase
    b. Sterilization
    c. Reduction
    d. Intensification

37. Based on information provided, why might other types of spraying plans be less desirable than backpack spraying?
    a. Human exposure to harmful chemicals
    b. The inability to target specific areas of mosquito proliferation
    c. Lack of community support for other methods
    d. Lack of household participation

38. In non-emergency situations, which factors of disease vector control can CASPERs be helpful in understanding?
    a. Community knowledge
    b. Lack of shelter
    c. Spraying support
    d. Risk assessment

39. Based on the passage, we can conclude that:
    a. Backpack spraying will become the primary disease control mechanism.
    b. The risks associated with mosquito-borne diseases can increase with hurricanes.
    c. CASPERs will become a primary tool used by government organizations.
    d. Disease vector problems will remain isolated to island nations and the developing world.

40. What is main takeaway of the final paragraph?
    a. Mosquito-borne diseases are a serious risk for most of the developing world.
    b. Community knowledge of mosquito abatement is well below where it should be.
    c. Community action on mosquito bite prevention needs to be supported by governments.
    d. CASPERs are a vital tool for assessing the public health of communities in emergency and non-emergency situations.

41. Based on the passage, what can we say about why mosquito-borne diseases are more prevalent post-hurricane?
    a. Environmental conditions are favorable.
    b. Mosquitoes don't bite during storms.
    c. Fewer preventative drugs are available.
    d. Residents have raised awareness.

42. Based on the passage, which statement would researchers most agree with?
    a. CASPERs are best used in emergency situations.
    b. Aerial spraying is most effective to combat Zika.
    c. Hurricanes greatly increase the negative risks associated with mosquitos in the US Virgin Islands.
    d. Bite prevention is the best strategy for fighting mosquito-borne diseases.

Science

Questions 43–52 are based on the following two passages:

Passage 1

In 1992, the City of Madison, Wisconsin, found concentrations of lead in their drinking water exceeding the 90th percentile action level of 0.015 mg/L set by EPA. Lead (Pb) is a naturally occurring metal that was commonly used in household plumbing materials, such as lead service lines and leaded solder joints, before limits were set on its use in 1986.

However, in houses built before 1986, lead pipes can still be in use. Lead is rarely found in source water, but it can enter tap water as the water enters pipes with lead in older systems. Since some homes have lead service lines, the water coming into the house may be transported via lead pipes even though there are no lead pipes inside the home. Brass plumbing fixtures can also contain small amounts of lead.

The Madison Water Utility chose to implement full lead service line replacement from 2001 through 2011 to eliminate the most significant source of lead in its water system. In 2003, sixty home taps were monitored after full lead service line replacement. They found that lead levels in the first liter of water were still high at some sites where the lead service line pipes had mostly been replaced within the previous four years. This phenomenon had been seen in other water systems, which had puzzled drinking water practitioners as to why elevated lead levels could persist for so long.

Coincident to the Madison Water Utility studies before and after the lead service line replacement program, lead service lines had been harvested from the water system and sent to EPA scientists. The EPA had the instrumentation and unique expertise to search for clues of lead release in pipe scales. That is, the materials that build up on the inside of pipes display chemical characteristics that reflect the chemical processes occurring in the water system, including the release of lead. The EPA conducted detailed analyses—color, texture, mineralogical and elemental composition—on five lead service pipe samples excavated between 2001 and 2006 from two different Madison neighborhoods.

Before the lead service line replacement program, Madison's water was delivered by an estimated 8000 lead service lines, which had been in service for 75 years or longer. The city's drinking water originated from numerous wells. The first set of lead service lines studied by EPA revealed that a highly insoluble and protective lead oxide compound had formed on the pipe walls. If all lead pipe walls had this formation, high lead releases would not be expected in the water system.

However, the second set of lead service lines came from a different neighborhood in the city. This neighborhood was fed by wells that were rich in manganese and iron. Both manganese and iron can form scales and accumulate metals, such as lead, from upstream sources, especially from upstream corroded lead pipes. EPA's results revealed that the accumulation of manganese and iron from the well water onto pipe walls had adsorbed lead and had the potential to crumble from the pipe walls and carry the lead to consumers' taps by means of the scale particulate matter entrained in the water. This finding corroborated with the results of the 2003 study where the higher lead concentration found at consumers' taps was mostly in particulate form. The presence of the manganese and iron scale on the pipe walls was the reason for high lead release in parts of the Madison water system, before and even after the lead pipes were removed.

As the 2003 residential study had shown, once the principal lead source was removed, it took more than four years in some cases for the accumulated lead to be released, which explains why lead levels remained high after the lead pipes had been replaced. Eventually, removing the source of lead did eliminate the significant lead concentration and achieved compliance with EPA's regulations.

Overall, this research showed that controlling lead exposure from water is more complicated than simply adding corrosion control chemicals to reduce the solubility of lead minerals. Buildup of manganese and iron scale in water pipes should also be considered as a source for accumulating and releasing lead, and other contaminants of concern, into water. What happened in Madison highlights the importance of analyzing pipe scales to understand how lead accumulates and releases into the water over time.

From: "Revealing the Complicated Nature of Tap Water Lead Contamination: A Madison, Wisconsin Case Study," July 30, 2018. EPA. https://www.epa.gov/sciencematters/revealing-complicated-nature-tap-water-lead-contamination-madison-wisconsin-case

Passage 2

An estimated six to ten million older homes across the country have lead service lines. Service lines connect individual houses to the water main in the street; this means that water coming into a house may be transported via a lead service line even if no lead pipes are visible inside the home. Lead can be transferred from the lead pipe into the drinking water when the pipe materials corrode, when there are physical disturbances to the pipe, or when there are changes to the quality of water entering the home.

Given that there are many lead service lines in use across the country, limiting corrosion is a necessary step to reduce potential lead exposure from drinking water. Public water systems can control corrosion through a variety of methods including strict control of key water quality parameters and proper addition of a phosphate or silicate corrosion control inhibitor. Public water systems sometimes use modeling to inform corrosion control. EPA researchers recently looked at how well these models were predicting what is happening in the real world.

Water systems in EPA's Region 5—comprised of Minnesota, Wisconsin, Michigan, Illinois, Indiana, and Ohio—shared lead service line pipes and water quality data with EPA researchers. Actual pieces of pipe were taken out of the water systems and sent to EPA where scientists examined the pipe scales, the materials that build up on the inside of pipes. These pipe scales reveal chemical characteristics that reflect the chemical and physical processes occurring within the water system including the release of lead into drinking water.

Once EPA researchers cut open the pipes and took the scales apart, they examined each layer of scale and the minerals that were present. Different minerals have different inherent solubilities which clued researchers in to which minerals may be dissolving into the water. EPA researchers looked at which minerals were predicted to form based on the modeling, and then looked at pipe scales found on the lead service lines from those systems to see which minerals really were forming.

EPA and other outside organizations have applied predictive solubility models to try and help systems pick the right corrosion control treatment that fits their system's individual needs. These models provide guidance regarding which mineral phases are predicted to control lead release in a given environment. EPA's model uses parameters like alkalinity and pH to predict which mineral would be expected to form in the lead service lines of a water system and how much dissolved lead you would expect to find in the water.

The researchers, including EPA's Jennifer Tully, Mike DeSantis, and Mike Schock, found various lead minerals, and other non-crystalline materials forming on the inside of the pipes. They discovered that there was almost always a mix of different lead minerals present in the scale. A little more than half of the lead service lines they looked at showed that the minerals present were not the same minerals that the models were predicting would be present.

Next, EPA researchers looked at water quality data from several systems that had supplied lead service lines for analysis. Since the pipe scales showed that the models were not always predicting the right mineral composition, the scientists wanted to see how well the corrosion control was working. What they found in limited sampling was that the models often have difficulty predicting real world scenarios due to the complexity of corrosion control, and this study shows the need for further evaluation and water sampling beyond just modeling to ensure that systems are using the correct treatment to keep lead out of drinking water.

From "EPA Researchers Help Water Systems Keep Lead out of Drinking Water," March 3, 2020.

43. Based on Passage 1, what complicates efforts to address lead exposure via household taps?
    a. Lack of reporting on the issue
    b. Lack of testing in many regions of the United States
    c. Inability to identify and track the sources of lead
    d. Multiple ways that lead accumulates and is released

44. Based on Passage 2, what is the public water system's role in lead contamination control?
    a. Informing residents of lead levels in their tap water
    b. Monitoring quality and corrosion control additives
    c. Replacing lead pipes in homes with safer plumbing systems
    d. Providing researchers the opportunity to track lead contamination sources

45. According to the findings in Passage 1, even if there are no lead pipes in the home, why is lead present?
    a. Some lead levels are present in all water
    b. Lead is present in the city's water source
    c. Testing is sometimes unreliable
    d. Lead service lines and accumulation leading to the home

46. According to the findings in Passage 2, why is modeling sometimes ineffective?
    a. Real-world scenarios don't always match the models.
    b. Researchers rarely have all the data to predict source points.
    c. Models are not updated to reflect new mineral presence.
    d. Changes in service lines change the outcome.

47. Based on the two passages, what would the authors most agree on?
    a. Low lead levels in water after line changes are not as concerning.
    b. Corrosion inhibitors in water source points are the most effective option.
    c. Lead is leaching from scaling in older pipes in homes.
    d. Well water is safer than municipal water.

48. In Passage 1, why were lead levels still present even after the primary lead source had been removed?
    a. Accumulated lead and minerals in the system took four years to resolve.
    b. The lead was determined to be coming from the water source itself.
    c. Corrosion inhibitors had not yet been introduced.
    d. False positive tests were reported due to other minerals.

49. In the fourth paragraph of Passage 2, what does the word "solubilities" mean?
    a. Concentration levels
    b. Bonding strengths
    c. Levels of condensation
    d. Abilities to dissolve

50. Based on both passages, in older homes, what initial mitigation strategy would the researchers agree on?
    a. High-end water filtration systems
    b. Replacing lead service lines
    c. Switching water sources
    d. Corrosion additives in municipal water

51. Which statement would both research groups agree with?
    a. Replacing lead lines in homes will resolve lead levels.
    b. Addressing scaling and mineral accumulation is crucial.
    c. Scientific modeling will continue to be a fundamental tool.
    d. Source point lead pollution is the biggest concern.

52. Which statement would researchers from both passages most DISAGREE with?
    a. Low lead levels can be handled by sink filtration.
    b. Lead eradication is easier with well water.
    c. Corrosion control at water sources is our best weapon.
    d. In five years, we can control lead levels in water.

Writing and Language Test

Read the essay entitled "Education is Essential to Civilization" and answer Questions 1–15.

Early in my career, (1) <u>a master's teacher shared this thought with me "Education is the last bastion of civility."</u> While I did not completely understand the scope of those words at the time, I have since come to realize the depth, breadth, truth, and significance of what he said. (2) <u>Education provides</u> society with a vehicle for (3) <u>raising it's children to be</u> civil, decent, human beings with something valuable to contribute to the world. It is really what makes us human and what (4) <u>distinguishes</u> us as <u>civilised</u> <u>creatures.</u>

Being "civilized" humans means being "whole" humans. Education must address the mind, body, and soul of students. (5) It would be detrimental to society, only meeting the needs of the mind, if our schools were myopic in their focus. As humans, we are multi-dimensional, multi-faceted beings who need more than head knowledge to survive. (6) The human heart and psyche have to be fed in order for the mind to develop properly, and the body must be maintained and exercised to help fuel the working of the brain. Education is a basic human right, and it allows us to sustain a democratic society in which participation is fundamental to its success. It should inspire students to seek better solutions to world problems and to dream of a more equitable society. Education should never discriminate on any basis, and it should create individuals who are self-sufficient, patriotic, and tolerant of (7) others' ideas.

(8) All children can learn. Although not all children learn in the same manner. All children learn best, however, when their basic physical needs are met and they feel safe, secure, and loved. Students are much more responsive to a teacher who values them and shows them respect as individual people. Teachers must model at all times the way they expect students to treat them and their peers. If teachers set high expectations for (9) there students, the students will rise to that high level. Teachers must make the well-being of students their primary focus and must not be afraid to let students learn from their own mistakes.

In the modern age of technology, a teacher's focus is no longer the "what" of the content, (10) but more importantly, the 'why.' Students are bombarded with information and have access to ANY information they need right at their fingertips. Teachers have to work harder than ever before to help students identify salient information (11) so to think critically about the information they encounter. Students have to (12) read between the lines, identify bias, and determine who they can trust in the milieu of ads, data, and texts presented to them.

Schools must work in consort with families in this important mission. While children spend most of their time in school, they are dramatically and indelibly shaped (13) with the influences of their family and culture. Teachers must not only respect this fact, (14) but must strive to include parents in the education of their children and must work to keep parents informed of progress and problems. Communication between classroom and home is essential for a child's success.

Humans have always aspired to be more, do more, and to better ourselves and our communities. This is where education lies, right at the heart of humanity's desire to be all that we can be. Education helps us strive for higher goals and better treatment of ourselves and others. I shudder to think what would become of us if education ceased to be the "last bastion of civility." (15) We must be unapologetic about expecting excellence from our students? Our very existence depends upon it.

1. What edit is needed to correct sentence 1 (reproduced below)?

Early in my career, (1) a master's teacher shared this thought with me "Education is the last bastion of civility."

a. NO CHANGE
b. a master's teacher shared this thought with me: "Education is the last bastion of civility."
c. a master's teacher shared this thought with me: "Education is the last bastion of civility".
d. a master's teacher shared this thought with me. "Education is the last bastion of civility."

2. What edit is needed to correct sentence 2 (reproduced below)?

(2) <u>Education provides</u> society with a vehicle for raising it's children to be civil, decent, human beings with something valuable to contribute to the world.

   a. NO CHANGE
   b. Education provide
   c. Education will provide
   d. Education providing

3. What edit is needed to correct sentence 3 (reproduced below)?

Education provides society with a vehicle for (3) <u>raising it's children to be</u> civil, decent, human beings with something valuable to contribute to the world.

   a. NO CHANGE
   b. raises its children to be
   c. raising its' children to be
   d. raising its children to be

4. Which of these, if any, is misspelled?
   a. None of these are misspelled.
   b. distinguishes
   c. civilised
   d. creatures

5. What edit is needed to correct sentence 5 (reproduced below)?

(5) <u>It would be detrimental to society, only meeting the needs of the mind, if our schools were myopic in their focus.</u>

   a. NO CHANGE
   b. It would be detrimental to society if our schools were myopic in their focus, only meeting the needs of the mind.
   c. Only meeting the needs of our mind, our schools were myopic in their focus, detrimental to society.
   d. Myopic is the focus of our schools, being detrimental to society for only meeting the needs of the mind.

6. Which of these sentences, if any, should begin a new paragraph?
   a. NO CHANGE
   b. The human heart and psyche have to be fed in order for the mind to develop properly, and the body must be maintained and exercised to help fuel the working of the brain.
   c. Education is a basic human right, and it allows us to sustain a democratic society in which participation is fundamental to its success.
   d. It should inspire students to seek better solutions to world problems and to dream of a more equitable society.

7. What edit is needed to correct sentence 7 (reproduced below)?

Education should never discriminate on any basis, and it should create individuals who are self-sufficient, patriotic, and tolerant of (7) others' ideas.

a. NO CHANGE
b. other's ideas
c. others ideas
d. others's ideas

8. What edit is needed to correct sentence 8 (reproduced below)?

(8) All children can learn. Although not all children learn in the same manner.

a. NO CHANGE
b. All children can learn although not all children learn in the same manner.
c. All children can learn although, not all children learn in the same manner.
d. All children can learn, although not all children learn in the same manner.

9. What edit is needed to correct sentence 9 (reproduced below)?

If teachers set high expectations for (9) there students, the students will rise to that high level.

a. NO CHANGE
b. they're students
c. their students
d. thare students

10. What edit is needed to correct sentence 10 (reproduced below)?

In the modern age of technology, a teacher's focus is no longer the "what" of the content, (10) but more importantly, the 'why.'

a. NO CHANGE
b. but more importantly, the "why."
c. but more importantly, the 'why'.
d. but more importantly, the "why".

11. What edit is needed to correct sentence 11 (reproduced below)?

Teachers have to work harder than ever before to help students identify salient information (11) so to think critically about the information they encounter.

a. NO CHANGE
b. and to think critically
c. but to think critically
d. nor to think critically

12. What edit is needed to correct sentence 12 (reproduced below)?

Students have to (12) <u>read between the lines, identify bias, and determine</u> who they can trust in the milieu of ads, data, and texts presented to them.

a. NO CHANGE
b. read between the lines, identify bias, and determining
c. read between the lines, identifying bias, and determining
d. reads between the lines, identifies bias, and determines

13. What edit is needed to correct sentence 13 (reproduced below)?

While children spend most of their time in school, they are dramatically and indelibly shaped (13) <u>with the influences</u> of their family and culture.

a. NO CHANGE
b. for the influences
c. to the influences
d. by the influences

14. What edit is needed to correct sentence 14 (reproduced below)?

Teachers must not only respect this fact, (14) <u>but must strive</u> to include parents in the education of their children and must work to keep parents informed of progress and problems.

a. NO CHANGE
b. but to strive
c. but striving
d. but strived

15. What edit is needed to correct sentence 15 (reproduced below)?

(15) <u>We must be unapologetic about expecting excellence from our students? Our very existence depends upon it.</u>

a. NO CHANGE
b. We must be unapologetic about expecting excellence from our students, our very existence depends upon it.
c. We must be unapologetic about expecting excellence from our students—our very existence depends upon it.
d. We must be unapologetic about expecting excellence from our students our very existence depends upon it.

Questions 16–24 are based on the following passage about Frankenstein *by Mary Shelley:*

(16) <u>One of the icon's of romantic and science fiction literature</u> remains Mary Shelley's classic, *Frankenstein, or The Modern Prometheus*. Schools throughout the world still teach the book in literature and philosophy courses. Scientific communities also engage in discussion on the novel. But why? Besides the novel's engaging (17) <u>writing style the story's central theme</u> remains highly relevant in a world of constant discovery and moral dilemmas. Central to the core narrative is the (18) <u>struggle between enlightenment and the cost of overusing power.</u>

The subtitle, *The Modern Prometheus*, encapsulates the inner theme of the story more than the main title of *Frankenstein*. As with many romantic writers, Shelley invokes the classical myths and (19) symbolism of Ancient Greece and Rome to high light core ideas. Looking deeper into the myth of Prometheus sheds light not only on the character of Frankenstein (20) but also poses a psychological dilemma to the audience. Prometheus is the titan who gave fire to mankind. (21) However, more than just fire he gave people knowledge and power. The power of fire advanced civilization. Yet, for giving fire to man, Prometheus is (22) punished by the gods bound to a rock and tormented for his act. This is clearly a parallel to Frankenstein—he is the modern Prometheus.

Frankenstein's quest for knowledge becomes an obsession. It leads him to literally create new life, breaking the bounds of conceivable science to illustrate that man can create life out of nothing. (23) Yet he ultimately faltered as a creator, abandoning his progeny in horror of what he created. Frankenstein then suffers his creature's wrath, (24) the result of his pride, obsession for power and lack of responsibility.

Shelley isn't condemning scientific achievement. Rather, her writing reflects that science and discovery are good things, but, like all power, it must be used wisely. The text alludes to the message that one must have reverence for nature and be mindful of the potential consequences. Frankenstein did not take responsibility or even consider how his actions would affect others. His scientific brilliance ultimately led to suffering.

16. Which of the following would be the best choice for this sentence (reproduced below)?

(16) One of the icon's of romantic and science fiction literature remains Mary Shelley's classic, *Frankenstein, or The Modern Prometheus*.

a. NO CHANGE
b. One of the icons of romantic and science fiction literature
c. One of the icon's of romantic, and science fiction literature,
d. The icon of romantic and science fiction literature

17. Which of the following would be the best choice for this sentence (reproduced below)?

Besides the novel's engaging (17) writing style the story's central theme remains highly relevant in a world of constant discovery and moral dilemmas.

a. NO CHANGE
b. writing style the central theme of the story
c. writing style, the story's central theme
d. the story's central theme's writing style

18. Which of the following would be the best choice for this sentence (reproduced below)?

Central to the core narrative is the (18) struggle between enlightenment and the cost of overusing power.

a. NO CHANGE
b. struggle between enlighten and the cost of overusing power.
c. struggle between enlightenment's cost of overusing power.
d. struggle between enlightening and the cost of overusing power.

19. Which of the following would be the best choice for this sentence (reproduced below)?

As with many romantic writers, Shelley invokes the classical myths and (19) symbolism of Ancient Greece and Rome to high light core ideas.

a. NO CHANGE
b. symbolism of Ancient Greece and Rome to highlight core ideas.
c. symbolism of ancient Greece and Rome to highlight core ideas.
d. symbolism of Ancient Greece and Rome highlighting core ideas.

20. Which of the following would be the best choice for this sentence (reproduced below)?

Looking deeper into the myth of Prometheus sheds light not only on the character of Frankenstein (20) but also poses a psychological dilemma to the audience.

a. NO CHANGE
b. but also poses a psychological dilemma with the audience.
c. but also poses a psychological dilemma for the audience.
d. but also poses a psychological dilemma there before the audience.

21. Which of the following would be the best choice for this sentence (reproduced below)?

(21) However, more than just fire he gave people knowledge and power.

a. NO CHANGE
b. However, more than just fire he gave people, knowledge, and power.
c. However, more than just fire, he gave people knowledge and power.
d. Besides actual fire, Prometheus gave people knowledge and power.

22. Which of the following would be the best choice for this sentence (reproduced below)?

Yet, for giving fire to man, Prometheus is (22) punished by the gods bound to a rock and tormented for his act.

a. NO CHANGE
b. punished by the gods, bound to a rock and tormented for his act.
c. bound to a rock and tormented as punishment by the gods.
d. punished for his act by being bound to a rock and tormented as punishment from the gods.

23. Which of the following would be the best choice for this sentence (reproduced below)?

   (23) <u>Yet he ultimately faltered as a creator,</u> abandoning his progeny in horror of what he created.

   a. NO CHANGE
   b. Yet, he ultimately falters as a creator by
   c. Yet, he ultimately faltered as a creator,
   d. Yet he ultimately falters as a creator by

24. Which of the following would be the best choice for this sentence (reproduced below)?
   Frankenstein then suffers his creature's wrath, (24) <u>the result of his pride, obsession for power and</u>
   <u>lack of responsibility.</u>

   a. NO CHANGE
   b. the result of his pride, obsession for power and lacking of responsibility.
   c. the result of his pride, obsession for power, and lack of responsibility.
   d. the result of his pride and also his obsession for power and lack of responsibility.

Questions 25–33 are based on the following passage:

   The power of legends continues to enthrall our imagination, provoking us both to wonder and
   explore. (25) <u>Who doesnt love a good legend?</u> Some say legends never (26) <u>die and this is</u>
   <u>certainly the case</u> for the most legendary creature of all, Bigfoot. To this day, people still claim
   sightings of the illusive cryptid. Many think of Bigfoot as America's monster, yet many nations
   have legends of a similar creature. In my own research I have found that Australia has the
   Yowie, China has the Yerin, and Russia has the Almas. (27) <u>Their all over the world, the</u> bigfoots
   and the legends tied to them. Does this mean they could exist?

   There are many things to consider when addressing (28) <u>this question but the chief factor</u> is
   whether there is credible evidence. (29) <u>For science to formally recognize that such a species</u>
   <u>exists, there needs to be physical proof.</u> While people have found supposed footprints and even
   (30) <u>captured photos and film of the creature, this validity of such evidence is up for debate.</u>
   There is room for uncertainty. Most visual evidence is out of focus, thus (31) <u>there is often</u>
   <u>skepticism whether such images are real.</u> Some researchers have even claimed to have hair and
   blood samples, but still there is doubt in the scientific community. The reason is simple: there
   needs to be a body or living specimen found and actively studied in order to prove the Bigfoots'
   existence.

   Yet, one cannot ignore the fact that (32) <u>hundreds of witnesses continuing to describe a</u>
   <u>creature</u> with uniform features all over the world. These bigfoot sightings aren't a modern
   occurrence either. Ancient civilizations have reported (33) <u>seeing Bigfoot as well including</u>
   <u>Native Americans.</u> It is from Native Americans that we gained the popular term Sasquatch,
   which is the primary name for the North American bigfoot. How does their testimony factor in?
   If indigenous people saw these animals, could they not have existed at some point? After all,
   when Europeans first arrived in Africa, they disbelieved the native accounts of the gorilla. But
   sure enough, Europeans eventually found gorillas and collected a body.

25. Which of the following would be the best choice for this sentence (reproduced below)?

(25) Who doesnt love a good legend?

a. NO CHANGE
b. Who does not love a good legend?
c. A good legend, who doesn't love one?
d. Who doesn't love a good legend?

26. Which of the following would be the best choice for this sentence (reproduced below)?

Some say legends never (26) die and this is certainly the case for the most legendary creature of all, Bigfoot.

a. NO CHANGE
b. die, and this is certainly the case
c. die; this is certainly the case
d. die. This is certainly the case

27. Which of the following would be the best choice for this sentence (reproduced below)?

(27) Their all over the world, the bigfoots and the legends tied to them.

a. NO CHANGE
b. There all over the world, the
c. They're all over the world, the
d. All over the world they are, the

28. Which of the following would be the best choice for this sentence (reproduced below)?

There are many things to consider when addressing (28) this question but the chief factor is whether there is credible evidence.

a. NO CHANGE
b. this question, but the chief factor
c. this question however the chief factor
d. this question; but the chief factor

29. Which of the following would be the best choice for this sentence (reproduced below)?

(29) For science to formally recognize that such a species exists, there needs to be physical proof.

a. NO CHANGE
b. Physical proof are needed in order for science to formally recognize that such a species exists.
c. For science to formally recognize that such a species exists there needs to be physical proof.
d. For science, to formally recognize that such a species exists, there needs to be physical proof.

30. Which of the following would be the best choice for this sentence (reproduced below)?

While people have found supposed footprints and even (30) captured photos and film of the creature, this validity of such evidence is up for debate.

a. NO CHANGE
b. captured photos and film of the creature. This validity of such evidence is up for debate.
c. captured photos and film of the creature, the validities of such evidence is up for debate.
d. captured photos and film of the creature, the validity of such evidence is up for debate.

31. Which of the following would be the best choice for this sentence (reproduced below)?

Most visual evidence is out of focus, thus (31) there is often skepticism whether such images are real.

a. NO CHANGE
b. often skepticism whether such images are real.
c. there is often skepticism, whether such images are real.
d. there is often skepticism weather such images are real.

32. Which of the following would be the best choice for this sentence (reproduced below)?

Yet, one cannot ignore the fact that (32) hundreds of witnesses continuing to describe a creature with uniform features all over the world.

a. NO CHANGE
b. hundreds of witnesses continuing to describing a creature
c. hundreds of witnesses continue to describe a creature
d. hundreds of the witnesses continue to described a creature

33. Which of the following would be the best choice for this sentence (reproduced below)?

Ancient civilizations have reported (33) seeing Bigfoot as well including Native Americans.

a. NO CHANGE
b. seeing Bigfoot, Native Americans as well.
c. seeing Bigfoot also the Native Americans.
d. seeing Bigfoot, including Native Americans.

Questions 34–40 are based on the following passage:

I have to admit that when my father bought a recreational vehicle (RV), I thought he was making a huge mistake. I didn't really know anything about RVs, but I knew that my dad was as big a "city slicker" as there was. (34) In fact, I even thought he might have gone a little bit crazy. On trips to the beach, he preferred to swim at the pool, and whenever he went hiking, he avoided touching any plants for fear that they might be poison ivy. Why would this man, with an almost irrational fear of the outdoors, want a 40-foot camping behemoth?

(35) The RV was a great purchase for our family and brought us all closer together. Every morning (36) we would wake up, eat breakfast, and broke camp. We laughed at our own

comical attempts to back The Beast into spaces that seemed impossibly small. (37) <u>We rejoiced</u> <u>as "hackers."</u> When things inevitably went wrong and we couldn't solve the problems on our own, we discovered the incredible helpfulness and friendliness of the RV community. (38) <u>We</u> <u>even made some new friends in the process.</u>

(39) <u>Above all, it allowed us to share adventures. While traveling across America,</u> which we could not have experienced in cars and hotels. Enjoying a campfire on a chilly summer evening with the mountains of Glacier National Park in the background or waking up early in the morning to see the sun rising over the distant spires of Arches National Park are memories that will always stay with me and our entire family. (40) <u>Those are also memories that my siblings</u> <u>and me</u> have now shared with our own children.

34. Which of the following would be the best choice for this sentence (reproduced below)?

    (34) <u>In fact, I even thought he might have gone a little bit crazy.</u>

    a. NO CHANGE
    b. Move the sentence so that it comes before the preceding sentence.
    c. Move the sentence to the end of the first paragraph.
    d. Omit the sentence.

35. In context, which is the best version of the underlined portion of this sentence (reproduced below)?

    (35) <u>The RV</u> was a great purchase for our family and brought us all closer together.

    a. NO CHANGE
    b. Not surprisingly, the RV
    c. Furthermore, the RV
    d. As it turns out, the RV

36. Which is the best version of the underlined portion of this sentence (reproduced below)?

    Every morning (36) <u>we would wake up, eat breakfast, and broke camp.</u>

    a. NO CHANGE
    b. we would wake up, eat breakfast, and break camp.
    c. would we wake up, eat breakfast, and break camp?
    d. we are waking up, eating breakfast, and breaking camp.

37. Which is the best version of the underlined portion of this sentence (reproduced below)?

    (37) <u>We rejoiced as "hackers."</u>

    a. NO CHANGE
    b. To a nagging problem of technology, we rejoiced as "hackers."
    c. We rejoiced when we figured out how to "hack" a solution to a nagging technological problem.
    d. To "hack" our way to a solution, we had to rejoice.

38. Which is the best version of the underlined portion of this sentence (reproduced below)?

(38) We even made some new friends in the process.

a. NO CHANGE
b. In the process was the friends we were making.
c. We are even making some new friends in the process.
d. We will make new friends in the process.

39. Which is the best version of the underlined portion of this sentence (reproduced below)?

(39) Above all, it allowed us to share adventures. While traveling across America, which we could not have experienced in cars and hotels.

a. NO CHANGE
b. Above all, it allowed us to share adventures while traveling across America
c. Above all, it allowed us to share adventures; while traveling across America
d. Above all, it allowed us to share adventures—while traveling across America

40. Which is the best version of the underlined portion of this sentence (reproduced below)?

(40) Those are also memories that my siblings and me have now shared with our own children.

a. NO CHANGE
b. Those are also memories that me and my siblings
c. Those are also memories that my siblings and I
d. Those are also memories that I and my siblings

Questions 41–45 are based on the following passage:

> We live in a savage world; that's just a simple fact. It is a time of violence, when the need for self-defense is imperative. (41) Martial arts, like Jiu-Jitsu, still play a vital role in ones survival. (42) Jiu-Jitsu, however doesn't justify kicking people around, even when being harassed or attacked. Today, laws prohibit the (43) use of unnecessary force in self-defense; these serve to eliminate beating someone to a pulp once they have been neutralized. Such laws are needed. Apart from being unnecessary to continually strike a person when (44) their down, its immoral. Such over-aggressive retaliation turns the innocent into the aggressor. Jiu-Jitsu provides a way for defending oneself while maintaining the philosophy of restraint and self-discipline. (45) Integrated into its core philosophy, Jiu-Jitsu tempers the potential to do great physical harm with respect for that power and for life.

41. Which of the following would be the best choice for this sentence (reproduced below)?

(41) Martial arts, like Jiu-Jitsu, still play a vital role in ones survival.

a. NO CHANGE
b. Martial arts, like Jiu-Jitsu, still play a vital role in one's survival.
c. Martial arts, like Jiu-Jitsu still play a vital role in ones survival.
d. Martial arts, like Jiu-Jitsu, still plays a vital role in one's survival.

42. Which of the following would be the best choice for this sentence (reproduced below)?

(42) <u>Jiu-Jitsu, however doesn't justify kicking people around,</u> even when being harassed or attacked.

   a. NO CHANGE
   b. Jiu-Jitsu, however, isn't justified by kicking people around,
   c. However, Jiu-Jitsu doesn't justify kicking people around,
   d. Jiu-Jitsu however doesn't justify kicking people around,

43. Which of the following would be the best choice for this sentence (reproduced below)?

Today, laws prohibit the (43) <u>use of unnecessary force in self-defense; these serve to eliminate</u> beating someone to a pulp once they have been neutralized.

   a. NO CHANGE
   b. use of unnecessary force in self-defense serving to eliminate
   c. use of unnecessary force, in self-defense, these serve to eliminate
   d. use of unnecessary force. In self-defense, these serve to eliminate

44. Which of the following would be the best choice for this sentence (reproduced below)?

Apart from being unnecessary to continually strike a person when (44) <u>their down, its immoral.</u>

   a. NO CHANGE
   b. their down, it's immoral.
   c. they're down, its immoral.
   d. they're down, it's immoral.

45. Which of the following would be the best choice for this sentence (reproduced below)?

(45) <u>Integrated into its core philosophy,</u> Jiu-Jitsu tempers the potential to do great physical harm with respect for that power, and for life.

   a. NO CHANGE
   b. Integrated into its core philosophy
   c. Integrated into it's core philosophy
   d. Integrated into its' core philosophy,

Math Test

1. If a car can travel 300 miles in 4 hours, how far can it go in an hour and a half?
   a. 100 miles
   b. 112.5 miles
   c. 135.5 miles
   d. 150 miles

2. At the store, Jan spends $90 on apples and oranges. Apples cost $1 each and oranges cost $2 each. If Jan buys the same number of apples as oranges, how many oranges did she buy?
   a. 20
   b. 25
   c. 30
   d. 35

3. What is the volume of a box with rectangular sides 5 feet long, 6 feet wide, and 3 feet high?
   a. 60 cubic feet
   b. 75 cubic feet
   c. 90 cubic feet
   d. 14 cubic feet

4. A train traveling 50 miles per hour takes a trip lasting 3 hours. If a map has a scale of 1 inch per 10 miles, how many inches apart are the train's starting point and ending point on the map if it travelled in a straight line?
   a. 14
   b. 12
   c. 13
   d. 15

5. A traveler takes an hour to drive to a museum, spends 3 hours and 30 minutes there, and takes half an hour to drive home. What percentage of his or her time was spent driving?
   a. 15%
   b. 30%
   c. 40%
   d. 60%

6. A truck is carrying three cylindrical barrels. Their bases have a diameter of 2 feet, and they have a height of 3 feet. What is the total volume of the three barrels in cubic feet?
   a. 3π
   b. 9π
   c. 12π
   d. 15π

7. Greg buys a $10 lunch with 5% sales tax. He leaves a $2 tip after his bill. How much money does he spend?
   a. $12.50
   b. $12
   c. $13
   d. $13.25

8. Marty wishes to save $150 over a 4-day period. How much must Marty save each day on average?
   a. $37.50
   b. $35
   c. $45.50
   d. $41

9. Bernard can make $80 per day. If he needs to make $300 and only works full days, how many days will this take?

    a. 6
    b. 3
    c. 5
    d. 4

10. A couple buys a house for $150,000. They sell it for $165,000. By what percentage did the house's value increase?

    a. 10%
    b. 13%
    c. 15%
    d. 17%

11. A school has 15 teachers and 20 teaching assistants. They have 200 students. What is the ratio of faculty to students?

    a. 3:20
    b. 4:17
    c. 5:54
    d. 7:40

12. A map has a scale of 1 inch per 5 miles. A car can travel 60 miles per hour. If the distance from the start to the destination is 3 inches on the map, how long will it take the car to make the trip?

    a. 12 minutes
    b. 15 minutes
    c. 17 minutes
    d. 20 minutes

13. Taylor works two jobs. The first pays $20,000 per year. The second pays $10,000 per year. She donates 15% of her income to charity. How much does she donate each year?

    a. $4500
    b. $5000
    c. $5500
    d. $6000

14. A box with rectangular sides is 24 inches wide, 18 inches deep, and 12 inches high. What is the volume of the box in cubic feet?

    a. 2
    b. 3
    c. 4
    d. 5

15. Kristen purchases $100 worth of CDs and DVDs. The CDs cost $10 each and the DVDs cost $15. If she bought four DVDs, how many CDs did she buy?

    a. 5
    b. 6
    c. 3
    d. 4

16. If Sarah reads at an average rate of 21 pages in four nights, how long will it take her to read 140 pages?
    a. 6 nights
    b. 26 nights
    c. 8 nights
    d. 27 nights

17. Mom's car drove 72 miles in 90 minutes. There are 5280 feet per mile. How fast did she drive in feet per second?
    a. 0.8 feet per second
    b. 48.9 feet per second
    c. 0.009 feet per second
    d. 70.4 feet per second

18. This chart indicates how many sales of CDs, vinyl records, and MP3 downloads occurred over the last year. Approximately what percentage of the total sales was from CDs?

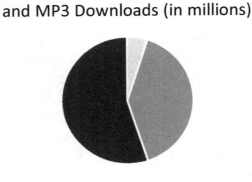

Total Sales of Vinyl Records, CDs, and MP3 Downloads (in millions)

Vinyl   ▪ CD   ▪ MP3

    a. 55%
    b. 25%
    c. 40%
    d. 5%

19. After a 20% sale discount, Frank purchased a new refrigerator for $850. How much did he save from the original price?
    a. $170
    b. $212.50
    c. $105.75
    d. $200

20. Which of the following is NOT a way to write 40 percent of N?

   a. $(0.4)N$

   b. $\frac{2}{5}N$

   c. $40N$

   d. $\frac{4N}{10}$

21. The graph of which function has an x-intercept of -2?

   a. $y = 2x - 3$
   b. $y = 4x + 2$
   c. $y = x^2 + 5x + 6$
   d. $y = -\frac{1}{2} \times 2^x$

22. The table below displays the number of three-year-olds at Kids First Daycare who are potty-trained and those who still wear diapers.

	Potty-trained	Wear diapers	Sum
Boys	26	22	48
Girls	34	18	52
Total	60	40	

What is the probability that a three-year-old girl chosen at random from the school is potty-trained?

   a. 52 percent
   b. 34 percent
   c. 65 percent
   d. 57 percent

23. A clothing company with a target market of U.S. boys surveys 2,000 twelve-year-old boys to find their height. The average height of the boys is 61 inches. For the above scenario, 61 inches represents which of the following?

   a. Sample statistic
   b. Population parameter
   c. Confidence interval
   d. Measurement error

24. A government agency is researching the average consumer cost of gasoline throughout the United States. Which data collection method would produce the most valid results?

   a. Randomly choosing one hundred gas stations in the state of New York
   b. Randomly choosing ten gas stations from each of the fifty states
   c. Randomly choosing five hundred gas stations from across all fifty states with the number chosen proportional to the population of the state
   d. All three methods would each produce equally valid results.

25. Suppose an investor deposits $1,200 into a bank account that accrues 1 percent interest per month. Assuming x represents the number of months since the deposit and y represents the money in the account, which of the following exponential functions models the scenario?

    a. $y = (0.01)(1200^x)$
    b. $y = (1200)(0.01^x)$
    c. $y = (1.01)(1200^x)$
    d. $y = (1200)(1.01^x)$

26. A student gets an 85% on a test with 20 questions. How many answers did the student solve correctly?

    a. 15
    b. 16
    c. 17
    d. 18

27. Four people split a bill. The first person pays for $\frac{1}{5}$, the second person pays for $\frac{1}{4}$, and the third person pays for $\frac{1}{3}$. What fraction of the bill does the fourth person pay?

    a. $\frac{13}{60}$

    b. $\frac{47}{60}$

    c. $\frac{1}{4}$

    d. $\frac{4}{15}$

28. Which of the following fractions is equal to 9.3?

    a. $9\frac{3}{7}$

    b. $\frac{903}{1000}$

    c. $\frac{9.03}{100}$

    d. $9\frac{3}{10}$

29. What is the solution to $3\frac{2}{3} - 1\frac{4}{5}$?

    a. $1\frac{13}{15}$

    b. $\frac{14}{15}$

    c. $2\frac{2}{3}$

    d. $\frac{4}{5}$

30. What is $\frac{420}{98}$ rounded to the nearest integer?

    a. 4
    b. 3
    c. 5
    d. 6

31. What is the solution to $4\frac{1}{3} + 3\frac{3}{4}$?

   a. $6\frac{5}{12}$

   b. $8\frac{1}{12}$

   c. $8\frac{2}{3}$

   d. $7\frac{7}{12}$

32. Five of six numbers have a sum of 25. The average of all six numbers is 6. What is the sixth number?

   a. 8

   b. 10

   c. 11

   d. 12

33. Suppose the function $y = \frac{1}{8}x^3 + 2x - 21$ approximates the population of a given city between the years 1900 and 2000 with x representing the year ($1900 = 0$) and y representing the population (in 1000s). Which of the following domains are relevant for the scenario?

   a. $(-\infty, \infty)$

   b. $[1900, 2000]$

   c. $[0, 100]$

   d. $[0, 0]$

34. What is the equation of a circle whose center is (0, 0) and whole radius is 5?

   a. $(x - 5)^2 + (y - 5)^2 = 25$

   b. $(x)^2 + (y)^2 = 5$

   c. $(x)^2 + (y)^2 = 25$

   d. $(x + 5)^2 + (y + 5)^2 = 25$

35. What is the equation of a circle whose center is (1, 5) and whole radius is 4?

   a. $(x - 1)^2 + (y - 25)^2 = 4$

   b. $(x - 1)^2 + (y - 25)^2 = 16$

   c. $(x + 1)^2 + (y + 5)^2 = 16$

   d. $(x - 1)^2 + (y - 5)^2 = 16$

36. Where does the point (-3, -4) lie on the circle with the equation $(x)^2 + (y)^2 = 25$?

   a. Inside of the circle.

   b. Outside of the circle.

   c. On the circle.

   d. There is not enough information to tell.

37. A drug needs to be stored at room temperature (68 °F). What is the equivalent temperature in degrees Celsius?

   a. 36 °C

   b. 72 °C

   c. 68 °C

   d. 20 °C

38. What is the slope of this line?

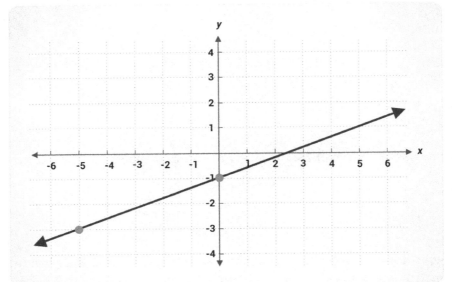

    a. 2

    b. $\frac{5}{2}$

    c. $\frac{1}{2}$

    d. $\frac{2}{5}$

39. What is the perimeter of the figure below? Note that the solid outer line is the perimeter.

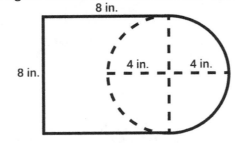

    a. 48.566 ft
    b. 36.566 ft
    c. 19.78 ft
    d. 30.566 ft

40. Which of the following equations best represents the problem below?
The width of a rectangle is 2 centimeters less than the length. If the perimeter of the rectangle is 44 centimeters, then what are the dimensions of the rectangle?
    a. $2l + 2(l - 2) = 44$
    b. $(l + 2) + (l + 2) + l = 48$
    c. $l \times (l - 2) = 44$
    d. $(l + 2) + (l + 2) + l = 44$

41. How will the following algebraic expression be simplified: $(5x^2 - 3x + 4) - (2x^2 - 7)$?
    a. x^5
    b. $3x^2 - 3x + 11$
    c. $3x^2 - 3x - 3$
    d. $x - 3$

42. What is 39% of 164?
    a. 63.96%
    b. 23.78%
    c. 6,396%
    d. 2.38%

43. Kimberley earns $10 an hour babysitting, and after 10 p.m., she earns $12 an hour, with the amount paid being rounded to the nearest hour accordingly. On her last job, she worked from 5:30 p.m. to 11 p.m. In total, how much did Kimberley earn for that job?
    a. $45
    b. $57
    c. $62
    d. $42

44. Keith's bakery had 252 customers go through its doors last week. This week, that number increased to 378. By what percentage did his customer volume increase?
    a. 26%
    b. 50%
    c. 35%
    d. 12%

No Calculator Questions

45. A family purchases a vehicle in 2005 for $20,000. In 2010, they decide to sell it for a newer model. They are able to sell the car for $8,000. By what percentage did the value of the family's car drop?
    a. 40%
    b. 68%
    c. 60%
    d. 33%

46. In May of 2010, a couple purchased a house for $100,000. In September of 2016, the couple sold the house for $93,000 so they could purchase a bigger one to start a family. How many months did they own the house?
    a. 76
    b. 54
    c. 85
    d. 93

47. At the beginning of the day, Xavier has 20 apples. At lunch, he meets his sister Emma and gives her half of his apples. After lunch, he stops by his neighbor Jim's house and gives him 6 of his apples. He then uses 3/4 of his remaining apples to make an apple pie for dessert at dinner. At the end of the day, how many apples does Xavier have left?
    a. 4
    b. 6
    c. 2
    d. 1

48. If $\frac{5}{2} \div \frac{1}{3} = n$, then n is between:
    a. 5 and 7
    b. 7 and 9
    c. 9 and 11
    d. 3 and 5

49. A closet is filled with red, blue, and green shirts. If $\frac{1}{3}$ of the shirts are green and $\frac{2}{5}$ are red, what fraction of the shirts are blue?
    a. $\frac{4}{15}$
    b. $\frac{1}{5}$
    c. $\frac{7}{15}$
    d. $\frac{1}{2}$

50. Shawna buys $2\frac{1}{2}$ gallons of paint. If she uses $\frac{1}{3}$ of it on the first day, how much does she have left?
    a. $1\frac{5}{6}$ gallons
    b. $1\frac{1}{2}$ gallons
    c. $1\frac{2}{3}$ gallons
    d. 2 gallons

51. What is the volume of a cylinder, in terms of π, with a radius of 6 centimeters and a height of 2 centimeters?
    a. 36π cm^3
    b. 24π cm^3
    c. 72π cm^3
    d. 48π cm^3

52. What is the length of the hypotenuse of a right triangle with one leg equal to 3 centimeters and the other leg equal to 4 centimeters?
    a. 7 cm
    b. 5 cm
    c. 25 cm
    d. 12 cm

53. Twenty is 40% of what number?

    a. 50

    b. 8

    c. 200

    d. 5000

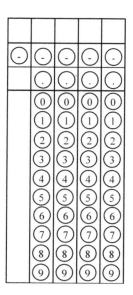

54. If Danny takes 48 minutes to walk 3 miles, how many minutes should it take him to walk 5 miles maintaining the same speed?

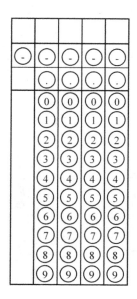

55. The perimeter of a 6-sided polygon is 56 cm. The lengths of three sides are 9 cm each. The lengths of two other sides are 8 cm each. What is the length of the missing side?

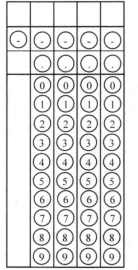

56. If the sine of $30° = x$, the cosine of what angle, in degrees, also equals x?

57. What is the value of $x^2 - 2xy + 2y^2$ when $x = 2$ and $y = 3$?

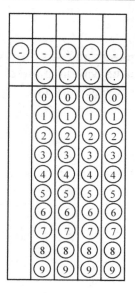

58. What is the value of x if $4x - 3 = 5$?

Essay Prompt

Bob's lawnmower shop has an excess supply of lawnmowers at the end of the summer season for the first time in years. Bob's son reasons that this excess supply should be heavily discounted and sold as quickly as possible to make room for winter equipment such as snowblowers and snowmobiles to maximize profits and revenue. Bob wants to keep the unsold lawnmowers and attempt to sell them again next year when demand returns.

Write a response in which you discuss what questions should be asked and resolved in order to validate Bob's strategy as a valid one versus his son's idea. Be sure to explain how easy those answers are to obtain, what assumptions they involve, and how the answers to the questions support Bob's strategy and confirm that it has a high probability of success.

Answer Explanations for Practice Test #2

Reading Test

1. C: Choice C is the correct answer. It is easy to imagine that she may have felt either anger (Choice A) or fear (Choice B) but those are not expressed in the text. Instead, while the passage mentions that her appearance was haughty (Choice D) it goes on to say that she likely felt agony, Choice C.

2. A: Choice A is correct. The passage does not mention feeling vengeful, Choice C. Further, the passage mentions that the crowd may have been tempted to ridicule and therefore be amused (Choice B) but they were not. Nor are they outraged (Choice D) as the passage notes that the nature of the punishment wouldn't lend itself to that feeling.

3. D: Choice D is correct. The passage states that Prynne was only required to stand on the scaffold, not be in the stockade, though the stockade was still in use. There is no mention of the stockade's not being used because she is a woman (Choice A) nor that the townspeople wanted her to be in it, Choice B. Finally, Choice C is incorrect because the passage does not discuss the stockade's use as a deterrent.

4. C: Choice C is correct. The passage clearly mentions that if the crowd had any inclination to ridicule, it would be suppressed by the presence of the governor, a judge, and other dignitaries. There is no mention in the passage of crimes committed by others, Choice A. Though her demeanor was proud, this was not the reason given for the lack of ridicule, so Choice B is not correct. Though there is mention of a death sentence, the passage does not say that the crowd wanted to see her put to death, so Choice D is also not correct.

5. B: Choice B is correct. Based on the context of the sentence and the paired adjective in the passage ("bitter"), a reader can infer that the meaning is not a positive one; therefore, Choice A is incorrect. Further, Choice C is incorrect as there is no logical reason why Prynne would be sympathetic. Finally, the text does not say that Prynne is mourning, so Choice D is incorrect.

6. A: Choice A is correct. The passage notes the agony felt by a prisoner who is exposed to spectators and laid bare for all to see. There is no mention of shackled ankles or walking speed, so Choice B and Choice C are incorrect. While there is mention of children later in the passage, it is not connected to the walk from prison to marketplace, so Choice D is incorrect.

7. D: Choice D is correct. The final paragraph of the passage notes that she expected venomous attacks, but the event was quiet instead. Therefore, Choice A which suggests cheering (Choice B) which suggests mocking, and Choice C, which suggests laughter, are all incorrect.

8. A: Choice A is correct. The passage's final sentences suggest the weight of her sentence and its looming duration as the reason for her need for emotional release. No one attacks her, so Choice B is incorrect. She is never placed in the stockade, so Choice C is incorrect. The crowd has been solemn and does not mock or ridicule her; while she might want to respond to the silence, that is not suggested by the passage. Choice D is, therefore, incorrect.

9. C: Choice C is correct. The passage opens with the third person narrator describing the scene, then moves to Prynne's experience, then that of the crowd and moves between those over the course of the paragraphs. While the judge is mentioned, an entire crowd acts as witness so Choice A is incorrect.

117

Prynne's husband is not present in this passage, so Choice *B* is incorrect. The town historian is not identified or discussed, though we do get town history, so Choice *D* is incorrect

10. A: Choice *A* is correct. The passage describes her at times as serene, strong, sustained, and fortified, so tenacious is the best fit. She walks almost proudly so Choice *B* and Choice *C* are incorrect. At the end of the passage, it notes that she wanted to yell, but it does not note that the desire is out of anger (Choice *D*) at the crowd or herself, but rather out of a need for release from the heaviness of the moment.

11. B: Choice *B* is correct. The writer notes specifically that they are tools to create a way of life we can enjoy. As such, while dictators do limit freedoms, but no in a way that populations typically enjoy, so Choice *A* is not correct. The passage does note that laws, but freedom ensures we can make our own, but not abuse or misuse so Choice *C* is incorrect. Finally, Choice *D* is incorrect because there is no mention of how we limit who uses the tool.

12. D: Choice *D* is correct. Roosevelt notes that without the processes and discussions that democracy requires, a dictatorship can achieve the same result much more quickly. While it is true that dictatorships may be benevolent, this is not the argument Roosevelt offers, so Choice *A* is incorrect. It is true that multiple factors enable dictators to rise to power, but this is not the reason Roosevelt offers, so Choice *B* is incorrect. Finally, Choice *C* suggests that dictators have been successful and popular, and though that might be true in some instances, it is also not what the passage discusses.

13. C: Choice *C* is correct. The passage states that antitrust laws maintain economic competition and prevent monopolies from limiting people's freedoms. Choice *A* is incorrect as antitrust laws are intended to protect consumers, and while they contribute to a relationship between governments and populations, that's not their primary goal. Choice *B* is incorrect as there are still ways in which businesses can manipulate markets, and so fairness isn't the ultimate outcome. Finally, businesses do compete with the government in certain industries, and that is not the purpose of antitrust laws, so Choice *D* is incorrect.

14: C: Choice *C* is correct. "Promulgated" means to promoted or spread. Therefore, Choices *A*, *B*, and *D* are incorrect as they are opposites and do not suggest spreading or championing an idea.

15: A: Choice *A* is correct. Multiple supporting points throughout the passage support Choice *A*, and paragraph 3 clarifies this main idea clearly. While Choice *B* (Choice *C*) and Choice *D* are all mentioned in the paragraph, they are mentioned as supporting points rather than the main idea.

16: B: Choice *B* is correct. While Choice *A* is true, that's not the reason to include it. The goal of mentioning the two is to compare the current situation to two major movements for democracy and their successful campaigns thereby highlighting the importance of such fights. While his audience might include both American and French citizens, he is speaking to a much larger audience, and that is not the primary reason, so Choice *C* is not the best answer. Finally, Choice *D* is incorrect as the date of the speech does not commemorate both events.

17. A: Choice *A* is correct. The speech highlights the ideals of human freedom and democracy, and altogether it implies that Roosevelt feels very strongly about these ideals. A satirical speech (Choice *B*) is one that scorns or ridicules, and that is incorrect. Choice *C* suggests someone is being attacked, and that's not accurate either. Finally, Choice *D* suggest she is appeasing someone, and that isn't correct either.

18. D: Choice *D* is correct. Because freedom and democracy are the main focus of each of the points Roosevelt brings up, it's clear that she values them above other ideals. While Choice *A* (Choice *B*) and Choice *C* might be true, they are not reflected as clearly in the passage as Choice *D* is. Those values are brought up in the passage, but only in support of freedom and democracy.

19. C: Choice *C* is correct. Roosevelt clearly states that her concern is that there is no guarantee a dictator will remain benevolent and no guarantee that power would be returned to the populace. While Choice *A* is true, it is not Roosevelt's concern. Choice *B* is also correct, but not discussed in the passage. Choice *D* is too extreme; Roosevelt does not say it's impossible for a dictator to be benevolent, though she does have concerns about them staying that way.

20. C: Choice *C* is correct. Roosevelt notes that secrecy ensures an individual's right to vote as she chooses. Choice *A* and Choice *D* may be true; public votes were known for their carnivalesque atmosphere, and private votes protect voters from violent confrontations with their political foes. However, these ideas are not stated in the passage. Choice *B* is incorrect as the opposite may be true. Privacy might increase the likelihood that, because the vote is secret, an individual can attest to voting one way but actually vote another without any accountability. This would make bribery a risk.

21. B: Choice *B* is correct. Johnson notes that Selma is an opportunity to right wrongs and seek justice as it relates to the denial of equal rights. Though a bridge played prominent in the Selma protest march, it was not the cause, so Choice *A* is incorrect. A man was killed, but that happened during the protest so could not be the cause. Therefore, Choice *C* is not correct. Finally, though Johnson discusses wealth in this passage, he does not say it is the cause of the events in Selma, so Choice *D* is incorrect.

22: C: Choice *C* is correct. Johnson says the cries created conversation and garnered the attention of the government to redress the issues raised. There is no mention of police response, so Choice *A* is incorrect. Similarly, there is no mention of the turnout at the protest, so Choice *B* is also incorrect. Finally, while Johnson does mention that one life was lost at the protest, he does not present this as cause for hope nor does he say it was "only" one life. Therefore, Choice *D* is also incorrect.

23. D: Choice *D* is correct. Johnson says that this challenge speaks "to the values, and the purposes, and the meaning of our beloved nation." There is no mention of other protests (Choice *A*), and there is no mention or whether those issues are easier (Choice *B*) or harder (Choice *C*) to solve, so those are not correct.

24. A: Choice *A* is correct. Johnson writes that a person's dignity "rests on his right to be treated as a man equal in opportunity to all others." Though Johnson discusses the protests (Choice *B*), American might and abundance (Choices *C* and *D*), they are not the defined as the source of a person's dignity.

25. A: Choice *A* is correct. What is righteous is morally just, and that characterizes Johnson's tone, so Choice *A* is the best answer. Choice *B*, "quixotic," means excessively or foolishly idealistic. To call Johnson's speech quixotic would mean that its goals of dignity and equality were foolish ones. Both scathing (Choice *C*) and vindictive (Choice *D*) suggest anger, harsh language, and bitterness. Johnson speaks of hope and progress, and he does not use language that suggests he is trying to chastise his audience, so these answers are incorrect.

26. D: Choice *D* is correct. Johnson says even if we "defeat every enemy" (Choice *C*), "double our wealth" (Choice *B*), or conquer the stars (Choice *A*), the greater failure is for a person to lose "his own soul." In this speech, "soul" refers to a person's moral or spiritual core, and to lose one's soul in this context means to lose moral or spiritual integrity.

27. C: Choice *C* is correct. Convocation means a gathering or a formal assembly. In this case, the people are summoning the leaders of government to an assembly. Based on context and conventional definitions of the words, Choices *A*, *B*, and *D* are incorrect.

28. C: Choice *C* is correct. The three quotations Johnson offers in this paragraph refer to the three ideals listed in Choice *C*. While there is mention of the military in this paragraph, it is not the purpose of the country, so Choice *A* is incorrect. While America is a world leader, it became a world leader some time after it was founded, so Choice *B* is incorrect. Finally, while America has prided itself on much, the goal was not to create envy, so Choice *D* is incorrect.

29. B: Choice *B* is correct. Johnson says these guardians "fought and died," and he says that they stand as guardians "around the world," referring to American troops. While some American citizens live in other parts of the world, they are not guarding liberty, so that answer is incorrect. "American government" (Choice *C*) is singular and "guardians" is plural, so that answer doesn't fit. The protestors are located in Selma and not "around the world," so Choice *D* is also incorrect.

30. D: Choice *D* is correct. Johnson is passionately speaking about civil rights and racial equality and is aiming to impart this mindset to all Americans.

31. B: Choice *B* is correct. Consent of the governed means the governed people must give permission to the government to lead them, presumably through elections and the structure of representation. Consent does not imply liking; therefore, Choice *A* is incorrect. Consent of the governed does not mean the governed must be treated fairly, so Choice *B* is incorrect. Consent of the governed requires that the governed have a voice; therefore, Choice *D* is incorrect.

32. A: Choice *A* is correct. According to the research, residents saw the risk of mosquito-borne illnesses as higher during the Zika outbreak despite the risk being higher post hurricane. It does not suggest that they saw no risk from either, so Choice *B* and Choice *C* are incorrect. Choice *D* is also incorrect as they did not see the risks as the same.

33. D: Choice *D* is correct. According to the research, one can infer that concerns about basic necessities like shelter, post-hurricane, may impact rates of mosquito-borne illnesses. There is no connection mentioned between Zika and hurricanes, though the research does compare the two in terms of mosquito problems, so Choice *A* is incorrect. Similarly, Choice *B* is incorrect as though the passage discusses resident concerns of risk and Zika, it does so in connection with mosquito-borne illnesses, not in regard to the spread of Zika specifically. Choice *C* is also incorrect because, again, though discussed together, there's no real correlation discussed in the passage.

34. C: Choice *C* is correct. The passage notes that CASPERs have been used to assess bite prevention and control-related factors, diseases, and breeding areas. CASPERs were used in the US Virgin Islands to mitigate mosquito-borne issues after the hurricane, not as a direct response to the hurricane, so Choice *A* is incorrect. While standing water might be one of the environmental factors included in post-hurricane mosquito proliferation, swamp size specifically was not addressed by CASPERs in the US Virgin Islands, so Choice *B* is incorrect. Backpack spraying was found to be an optimal strategy after evaluating information from CASPERs; the CASPER itself was not about backpack spraying, so Choice *D* is incorrect.

35. B: Choice *B* is correct. In the context of health, and as used in the passage, vector-related refers to the pathway by which a disease or pathogen is spread. For example, a vector-borne illness is transmitted from an insect or other organism (considered the vector) to another organism or plant.

36. C: Choice C is correct. In the third paragraph, the passage refers to the need for increased mosquito "abatement." Based on the focus of the paragraph, Choice A and Choice D would represent the spread of mosquitoes and are therefore incorrect. While sterilization might be an option for abatement, it's not mentioned specifically nor is it the only tactic for abatement, so Choice B is incorrect.

37. B: Choice B is correct. Based on the information provided, backpack spraying best addresses the need to spray targeted areas with greatest risk. Human exposure to mosquito bites is mentioned, but not to the chemicals used in spraying, so Choice A is incorrect. Though the passage mentions community support for spraying, it doesn't specify the type of support, so Choice C is incorrect. Finally, while the passage mentions household-level prevention, it doesn't specify how household participation affects spraying, so Choice D is also incorrect.

38. A: Choice A is correct. According to the passage, CASPERs may help assess community campaigns and reveal community knowledge, attitudes, and practices. Lack of shelter and spraying support are noted directly in reference to post-hurricane concerns, which is an emergency situation, so Choice B and Choice C are incorrect. While some of the efforts of CASPERs may reveal community risks, particularly as it relates to actual versus perceived risks, it is not listed among the non-emergency applications in the final paragraph and so Choice D is incorrect.

39. B: Choice B is correct. While backpack spraying is noted as the acceptable option, we cannot determine how communities will move forward with its inclusion in their response without more information, so Choice A is incorrect. Similarly, this passaged argues for the inclusion of CASPERs as a tool in emergency and non-emergency situations, but we cannot draw a conclusion regarding future use, so Choice C is incorrect. Given the nature of increased storm activity across the world, we cannot be certain that these disease vector problems will remain isolated to island nations, so Choice D is incorrect. However, given the increase in severe storms, we can conclude that risks with mosquito-borne illnesses will also increase, meaning Choice B is correct.

40. D: Choice D is correct. The primary focus of this paragraph is to discuss how CASPER can be used to assist communities and governments in addressing both emergency and non-emergency issues. While Choices A, B, and C are true, they are supporting details in the paragraph and not the focus of the passage itself.

41. A: Choice A is correct. The passage notes, in paragraph 1, that environmental conditions for mosquito breeding and exposure are more favorable after a hurricane. While it might be true that mosquitoes bite less during storms (Choice B), or that fewer preventative medications are available (Choice C), these are not discussed in the passage. Finally, it is noted that residents seem to be aware of an elevated risk post-hurricane, but awareness does not cause disease, so Choice D is incorrect.

42. C: Choice C is correct. The passage notes that CASPERs would be incredibly useful in non-emergency situations as well, so Choice A is incorrect. Aerial spraying is not mentioned specifically, especially in relation to Zika, so Choice B is incorrect. Finally, researchers agree that controlling mosquito-borne diseases involves bite prevention, but also requires control, community knowledge and other factors, so while Choice D is part of the solution, it may not be the best option.

43. D: Choice D is correct. Both Choice A and Choice B suggest that researchers and residents are unaware of the lead problem in much of America's water. However, the existence of the research itself contradicts both these answers. Similarly, Choice C directly contradicts the research presented in both passages, which not only identify but also track the lead sources.

44. B: Choice *B* is correct. The passage states that public systems can measure water quality and use corrosion inhibitors to improve water quality. While a city may inform residents of high lead levels, that's not noted in this passage regarding the public system's role. Therefore, Choice *A* is incorrect. Similarly, public water systems are not responsible for replacing home piping, Choice *C*. Further, while systems clearly allow and likely encourage research, their objective is to mitigate lead exposure, so Choice *D* is not the best answer.

45. D: Choice *D* is correct. The research suggests that if the lead pipes have been removed from the home, the city's service lines could be a source for lead. Choice *A* and Choice *C* are not discussed in the passage and are, therefore, incorrect. While Choice *B* is possible, it is not presented as the cause nor discussed in the passage.

46. A: Choice *A* is correct. While Choice *B*, Choice *C*, and Choice *D* are all reasonable possibilities, they are not discussed in the article. Though we know researchers do not have all the data (Choice *B*), there is no reference to this being related to source points. We do not get any information about how researchers are adapting the models to new information (Choice *C*), so that is incorrect as well. Finally, there is no data presented regarding how changes in the service lines impact the models being used.

47. C: Choice *C* is correct. There is no data to support Choice *A*; in fact, some of the research suggests that even changes the lines has little initial impact on lead levels. Choice *B* is mentioned as a mitigation and control factor, but it is not suggested that it's the most effective. Choice *D* is not discussed at all. In fact, lead levels are a concern regardless of the water source.

48. A: Choice *A* is correct. There was no mention in passage 1 of lead levels in the water sources themselves, so Choice *B* is incorrect. Similarly, passage 1 discusses corrosion inhibitors, but not as related to their introduction and its impact on the research, so Choice *C* is incorrect. There is no mention of false positives, so Choice *D* is incorrect.

49. D: Choice *D* is correct. Based on the sentence it is used in, one can infer the meaning from the context. Later in the sentence, the writer uses the word "dissolve" in place of "solubility." Choice *B* and Choice *C* are both the opposite of the correct meaning and, therefore, incorrect. Choice *A* is the result of the dissolution and also incorrect.

50. B: Choice *B* is correct. Both passages focus on the need to replace lead service lines in older homes as the first step to addressing lead levels in tap water. High-end water filtration systems are not discussed at all, so Choice *A* is incorrect. Lead levels were found in both types of water sources, so Choice *C* is not the correct answer. While Choice *D* is one of the methods discussed, it is secondary to the replacement of service lines and, therefore, incorrect.

51. B: Choice *B* is correct. In both passages, the researchers focus on the accumulation of minerals and scaling in pipes, problems that must be handled through service line replacement and corrosion inhibitors. While replacing lines in older homes, Choice *A*, is one solution, neither group believes it will resolve the issue. Choice *C* may be true, and though it is discussed in the second passage, it is not discussed in the first passage; nor does either passage discuss its role in future research. Finally, source point pollution is not discussed in either passage though water sources are discussed, so Choice *D* is incorrect.

52. D: Choice *D* is correct. Because of the scope of work, and because of the existence of lead in water up to four years after lead lines have been replaced in the home, one can infer that Choice *D* is the statement researchers would most disagree with. Because they do not discuss tap filtration, we cannot gauge where researchers would stand on how this method would work, so Choice A is incorrect. Well water is not without its own issues, particularly in relation to scaling and additional minerals. Researchers would likely agree that the issue is different rather than determining whether it's easier, so Choice B is incorrect. Finally, corrosion control (Choice C) is a valuable weapon, but researchers would not agree that it's the best option, based on both passages. Instead, they'd likely argue that corrosion control is part of a multi-step process.

Writing and Language Test

1. B: Choice B is correct. Here, a colon is used to introduce an explanation. Colons either introduce explanations or lists. Additionally, the quote ends with the punctuation inside the quotes, unlike Choice C.

2. A: The verb tense in this passage is predominantly in the present tense, so Choice A is the correct answer. Choice B is incorrect because the subject and verb do not agree. It should be "Education provides," not "Education provide." Choice C is incorrect because the passage is in present tense, and "Education will provide" is future tense. Choice D doesn't make sense when placed in the sentence.

3. D: The possessive form of the word "it" is "its." The contraction "it's" denotes "it is." Thus, Choice A is wrong. The word "raises" in Choice B makes the sentence grammatically incorrect. Choice C adds an apostrophe at the end of "its." While adding an apostrophe to most words would indicate possession, adding 's to the word "it" indicates a contraction.

4. C: The word *civilised* should be spelled *civilized*. The words *distinguishes* and *creatures* are both spelled correctly.

5. B: Choice B is correct because it provides clarity by describing what "myopic" means right after the word itself. Choice A is incorrect because the explanation of "myopic" comes before the word; thus, the meaning is skewed. It's possible that Choice C makes sense within context. However, it's not the *best* way to say this. Choice D is confusingly worded. Using "myopic focus" is not detrimental to society; however, the way D is worded makes it seem that way.

6. C: Again, we see where the second paragraph can be divided into two parts due to separate topics. The first section's main focus is education addressing the mind, body, and soul. The first section, then, could end with the concluding sentence, "The human heart and psyche . . ." The next sentence to start a new paragraph would be "Education is a basic human right." The rest of this paragraph talks about what education is and some of its characteristics.

7. A: Choice A is correct because the phrase "others' ideas" is both plural and indicates possession. Choice B is incorrect because "other's" indicates only one "other" that's in possession of "ideas," which is incorrect. Choice C is incorrect because no possession is indicated. Choice D is incorrect because the word *other* does not end in s. Others's is not a correct form of the word in any situation.

8. D: This sentence must have a comma before "although" because the word "although" is connecting two independent clauses. Thus, Choices B and C are incorrect. Choice A is incorrect because the second sentence in the underlined section is a fragment.

9. C: Choice C is the correct choice because the word "their" indicates possession, and the text is talking about "their students," or the students of someone. Choice A, "there," means at a certain place and is incorrect. Choice B, "they're," is a contraction and means "they are." Choice D is not a word.

10. B: Choice *B* uses all punctuation correctly in this sentence. In American English, single quotes should only be used if they are quotes within a quote, making Choices *A* and *C* incorrect. Additionally, punctuation here should go inside the quotes, making Choice *D* incorrect.

11. B: Choice *B* is correct because the conjunction *and* is used to connect phrases that are to be used jointly, such as teachers working hard to help students "identify salient information" and to "think critically." The conjunctions *so*, *but*, and *nor* are incorrect in the context of this sentence.

12. A: Choice *A* has consistent parallel structure with the verbs "read," "identify," and "determine." Choices *B* and *C* have faulty parallel structure with the words "determining" and "identifying." Choice *D* has incorrect subject/verb agreement. The sentence should read, "Students have to read . . . identify . . . and determine."

13. D: The correct choice for this sentence is that "they are . . . shaped by the influences." The prepositions "for," "to," and "with" do not make sense in this context. People are *shaped by*, not *shaped for, shaped to,* or *shaped with.*

14. A: To see which answer is correct, it might help to place the subject, "Teachers," near the verb. Choice *A* is correct: "Teachers . . . must strive" makes grammatical sense here. Choice *B* is incorrect because "Teachers . . . to strive" does not make grammatical sense. Choice *C* is incorrect because "Teachers must not only respect . . . but striving" eschews parallel structure. Choice *D* is incorrect because it is in past tense, and this passage is in present tense.

15. C: Choice *C* is correct because it uses an em-dash. Em-dashes are versatile. They can separate phrases that would otherwise be in parenthesis, or they can stand in for a colon. In this case, a colon would be another decent choice for this punctuation mark because the second sentence expands upon the first sentence. Choice *A* is incorrect because the statement is not a question. Choice *B* is incorrect because adding a comma here would create a comma splice. Choice *D* is incorrect because this creates a run-on sentence since the sentence contains two independent clauses.

16. B: Choice *B* is correct because it removes the apostrophe from *icon's*, since the noun *icon* is not possessing anything. This conveys the author's intent of setting *Frankenstein* apart from other icons of the romantic and science fiction genres. Choices *A* and *C* are therefore incorrect. Choice *D* is a good revision but alters the meaning of the sentence—*Frankenstein* is one of the icons, not the sole icon.

17 C: Choice *C* correctly adds a comma after *style*, successfully joining the dependent and the independent clauses as a single sentence. Choice *A* is incorrect because the dependent and independent clauses remain unsuccessfully combined without the comma. Choices *B* and *D* do nothing to fix this.

18. A: Choice *A* is correct, as the sentence doesn't require changes. Choice *B* incorrectly changes the noun *enlightenment* into the verb *enlighten*. Choices *C* and *D* alter the original meaning of the sentence.

19. B: Choice *B* is correct, fixing the incorrect split of *highlight*. This is a polyseme, a word combined from two unrelated words to make a new word. On their own, *high* and *light* make no sense for the sentence, making Choice *A* incorrect. Choice *C* incorrectly decapitalizes *Ancient*—since it modifies *Greece* and works with the noun to describe a civilization, *Ancient Greece* functions as a proper noun, which should be capitalized. Choice *D* uses *highlighting*, a gerund, but the present tense of *highlight* is what works with the rest of the sentence; to make this change, a comma would be needed after *Rome*.

20. A: Choice *A* is correct, as *not only* and *but also* are correlative pairs. In this sentence, *but* successfully transitions the first part into the second half, making punctuation unnecessary. Additionally, the use of *to* indicates that an idea or challenge is being presented to the reader. Choices *B*, *C*, and *D* are not as active, meaning these revisions weaken the sentence.

21. D: Choice *D* is correct, adding finer details to help the reader understand exactly what Prometheus did and his impact: fire came with knowledge and power. Choice *A* lacks a comma after *fire*. Choice *B* inserts unnecessary commas since *people* is not part of the list *knowledge and power*. Choice *C* is a strong revision but could be confusing, hinting that the fire was knowledge and power itself, as opposed to being symbolized by the fire.

22. C: Choice *C* reverses the order of the section, making the sentence more direct. Choice *A* lacks a comma after *gods*, and although Choice *B* adds this, the structure is too different from the first half of the sentence to flow correctly. Choice *D* is overly complicated and repetitious in its structure even though it doesn't need any punctuation.

23. B: Choice *B* fixes the two problems of the sentence, changing *faltered* to present tense in agreement with the rest of the passage, and correctly linking the two dependent clauses. Choice *A* is therefore incorrect. Choice *C* does not correct the past tense of *faltered*. Choice *D* correctly adds the conjunction *by*, but it lacks a comma after the conjunction *yet*.

24. C: Choice *C* successfully applies a comma after *power*, distinguishing the causes of Frankenstein's suffering and maintaining parallel structure. Choice *A* is thus incorrect. Choice *B* lacks the necessary punctuation and unnecessarily changes *lack* to a gerund. Choice *D* is unnecessarily wordy, making the sentence more cumbersome.

25. D: Choice *D* correctly inserts an apostrophe into the contraction *doesn't*. Choice *A* is incorrect because of this omission. Choices *B* and *C* are better than the original but do not fit well with the informal tone of the passage.

26. B: Choice *B* is correct, successfully combining the two independent clauses of this compound sentence by adding a comma before *and* to create the effective pause and transition between clauses. Choice *A* does not join the independent clauses correctly. Choices *C* and *D* offer alternate ways of joining these clauses, but since *and* is already part of the sentence, adding the comma is the most logical choice. This also keeps the informal tone set by the rest of the passage.

27. C: Choice *C* correctly fixes the homophone issue of *their* and *they're*. *Their* implies ownership, which is not needed here. The author intends *they're*, a contraction of *they are*. Thus, Choices *A* and *B* are incorrect because they use the homophone *there*. Choice *D* eliminates the homophone issue altogether, but the sentence becomes more clunky because of that.

28. B: Choice *B* correctly joins the two independent clauses with a comma before *but*. Choice *A* is incorrect because, without the comma, it is a run-on sentence. Choice *C* also lacks punctuation and uses *however*, which should be reserved for starting a new sentence or perhaps after a semicolon. Choice *D* is incorrect because the semicolon throws off the sentence structure and is incorrectly used; the correct revision would have also removed *but*.

29. A: Choice *A* is correct because the sentence does not require modification. Choice *B* is incorrect because it uses the faulty subject-verb agreement, "Physical proof are." Choice *C* is incorrect because a comma would need to follow *exists*. Choice *D* is incorrect because the comma after *science* is unnecessary.

30. D: Choice *D* correctly changes *this* to *the* and retains *validity*, making it the right choice. Choices *A* and *B* keep *this*, which is not as specific as *the*. Choice *C* incorrectly pluralizes *validity*.

31. A: Choice *A* is correct because the sentence is fine without revisions. Choice *B* is incorrect, since removing *there is* is unnecessary and confusing. Choice *C* is incorrect since it inserts an unnecessary comma. Choice *D* introduces a homophone issue: *weather* refers to climatic states and atmospheric events, while *whether* expresses doubt, which is the author's intent.

32. C: Choice *C* correctly changes *continuing* to the present tense. Choice *A* is incorrect because of this out-of-place gerund use. Choice *B* not only does not fix this issue but also incorrectly changes *describe* into a gerund. While Choice *D* correctly uses *continue*, *describe* is incorrectly put in the past tense.

33. D: Choice *D* is correct, since it eliminates the unnecessary *as well* and adds a comma to separate the given example, making the sentence more direct. Choice *A* seems repetitive with *as well*, since it has *including*, and at the least needs punctuation. Choice *B* is poorly constructed, taking out the clearer *including*. Choice *C* also makes little sense.

34. B: Move the sentence so that it comes before the preceding sentence. For this question, place the underlined sentence in each prospective choice's position. To keep it as-is is incorrect because the father "going crazy" doesn't logically follow the fact that he was a "city slicker." Choice *C* is incorrect because the sentence in question is not a concluding sentence and does not transition smoothly into the second paragraph. Choice *D* is incorrect because the sentence doesn't necessarily need to be omitted since it logically follows the very first sentence in the passage.

35. D: Choice *D* is correct because "As it turns out" indicates a contrast from the previous sentiment, that the RV was a great purchase. Choice *A* is incorrect because the sentence needs an effective transition from the paragraph before. Choice *B* is incorrect because the text indicates it *is* surprising that the RV was a great purchase because the author was skeptical beforehand. Choice *C* is incorrect because the transition "Furthermore" does not indicate a contrast.

36. B: This sentence calls for parallel structure. Choice *B* is correct because the verbs "wake," "eat," and "break" are consistent in tense and parts of speech. Choice *A* is incorrect because the words "wake" and "eat" are present tense while the word "broke" is in past tense. Choice *C* is incorrect because this turns the sentence into a question, which doesn't make sense within the context. Choice *D* is incorrect because it breaks tense with the rest of the passage. "Waking," "eating," and "breaking" are all present participles, and the context around the sentence is in past tense.

37. C: Choice *C* is correct because it is clear and fits within the context of the passage. Choice *A* is incorrect because "We rejoiced as 'hackers'" does not give a reason why hacking was rejoiced. Choice *B* is incorrect because it does not mention a solution being found and is therefore not specific enough. Choice *D* is incorrect because the meaning is eschewed by the helping verb "had to rejoice," and the sentence suggests that rejoicing was necessary to "hack" a solution.

38. A: The original sentence is correct because the verb tense as well as the order of the sentence makes sense in the given context. Choice *B* is incorrect because the order of the words makes the sentence more confusing than it otherwise would be. Choice *C* is incorrect because "We are even making" is in present tense. Choice *D* is incorrect because "We will make" is future tense. The surrounding text of the sentence is in past tense.

39. B: Choice *B* is correct because there is no punctuation needed if a dependent clause ("while traveling across America") is located behind the independent clause ("it allowed us to share adventures"). Choice *A* is incorrect because there are two dependent clauses connected and no independent clause, and a complete sentence requires at least one independent clause. Choice *C* is incorrect because of the same reason as Choice *A*. Semicolons connect closely related independent clauses on either side of the semicolon. Choice *D* is incorrect because the dash simply interrupts the complete sentence.

40. C: The rule for "me" and "I" is that one should use "I" when it is the subject pronoun of a sentence, and "me" when it is the object pronoun of the sentence. Break the sentence up to see if "I" or "me" should be used. To say "Those are memories that I have now shared" makes more sense than to say "Those are memories that me have now shared." Choice *D* is incorrect because "my siblings" should come before "I."

41. B: Choice *B* is correct because it adds an apostrophe to *ones*, which indicates *one's* possession of *survival*. Choice *A* doesn't do this, so it is incorrect. This is the same for Choice *C*, but that option also takes out the crucial comma after *Jiu-Jitsu*. Choice *D* is incorrect because it changes *play* to *plays*. This disagrees with the plural *martial arts*, which is given by having an example of its many forms, *Jiu-Jitsu*. Therefore, *play* is required.

42. C: Choice *C* is the best answer because it most clearly defines the point that the author is trying to make. The original sentence would need a comma after *however* in order to continue the sentence fluidly—but this option isn't available. Choice *B* is close, but this option changes the meaning of the sentence. Therefore, the best alternative is to begin the sentence with *however* and have a comma follow right after it in order to introduce a new idea. The original context is still maintained, but the flow of the language is more streamlined. Thus, Choice *A* is incorrect. Choice *D* would need a comma before and after *however*, so it is also incorrect.

43. A: Choice *A* is the best answer for several reasons. To begin, the section is grammatically correct in using a semicolon to connect the two independent clauses. This allows the two ideas to be connected without separating them. In this context, the semicolon makes more sense for the overall sentence structure and passage as a whole. Choice *B* is incorrect because it forms a run-on. Choice *C* applies a comma in incorrect positions. Choice *D* separates the sentence in a place that does not make sense for the context.

44. D: Choice *D* is the correct answer because it fixes two key issues. First, *their* is incorrectly used. *Their* is a possessive indefinite pronoun and also an antecedent—neither of these fit the context of the sentence, so Choices *A* and *B* are incorrect. What should be used instead is *they're*, which is the contraction of *they are*, emphasizing action or the result of action in this case. Choice *D* also corrects another contraction-related issue with *its*. Again, *its* indicates possession, while *it's* is the contraction of *it is*. The latter is what's needed for the sentence to make sense and be grammatically correct. Thus, Choice *C* is also incorrect.

45. A: Choice *A* is correct because the section contains no errors and clearly communicates the writer's point. Choice *B* is incorrect because it lacks a comma after *philosophy*, needed to link the first clause with the second. Choice *C* also has this issue but additionally alters *its* to *it's*; since *it is* does not make sense in this sentence, this is incorrect. Choice *D* is incorrect because *its* is already plural possessive and does not need an apostrophe on the end.

Math Test

1. B: 300 miles in 4 hours is $\frac{300}{4} = 75$ miles per hour. In 1.5 hours, the car will go 1.5×75 miles, or 112.5 miles.

2. C: One apple/orange pair costs $3 total. Therefore, Jan bought $\frac{90}{3} = 30$ total pairs, and hence, she bought 30 oranges.

3. C: The formula for the volume of a box with rectangular sides is the length times width times height, so $5 \times 6 \times 3 = 90$ cubic feet.

4. D: First, the train's journey in the real word is $3 \times 50 = 150$ miles. On the map, 1 inch corresponds to 10 miles, so there is $\frac{150}{10} = 15$ inches on the map.

5. B: The total trip time is $1 + 3.5 + 0.5 = 5$ hours. The total time driving is $1 + 0.5 = 1.5$ hours. So, the fraction of time spent driving is 1.5 / 5 or 3 / 10.

To get the percentage, convert this to a fraction out of 100. The numerator and denominator are multiplied by 10, with a result of 30 / 100. The percentage is the numerator in a fraction out of 100, so 30%.

6. B: The formula for the volume of a cylinder is $\pi r^2 h$, where r is the radius and h is the height. The diameter is twice the radius, so these barrels have a radius of 1 foot. That means each barrel has a volume of:

$$\pi \times 1^2 \times 3 = 3\pi \text{ cubic feet}$$

Since there are three of them, the total is $3 \times 3\pi = 9\pi$ cubic feet.

7. A: The tip is not taxed, so he pays 5% tax only on the $10. The tax is 5% of $10, or $0.05 \times 10 = \$0.50$. Add up $10 + $2 + $0.50 to get $12.50.

8. A: The first step is to divide up $150 into four equal parts. $150 \div 4$ is 37.5, so she needs to save an average of $37.50 per day.

9. D: The number of days can be found by taking the total amount Bernard needs to make and dividing it by the amount he earns per day:

$$\frac{300}{80} = \frac{30}{8} = \frac{15}{4} = 3.75$$

But Bernard is only working full days, so he will need to work 4 days, since 3 days is not a sufficient amount of time.

10. A: The value went up by $165,000 - $150,000 = $15,000.

Out of $150,000, this is $\frac{15,000}{150,000} = \frac{1}{10}$.

Convert this to having a denominator of 100, which yields a result of $\frac{10}{100}$ or 10%.

11. D: The total faculty is $15 + 20 = 35$. Therefore, the faculty to student ratio is 35:200. Then, to simplify this ratio, both the numerator and the denominator are divided by 5, since 5 is a common factor of both, which yields 7:40.

12. B: The journey will be $5 \times 3 = 15$ miles. A car traveling at 60 miles per hour is traveling at 1 mile per minute. The resulting equation would be:

$$\frac{15 \text{ mi}}{1 \frac{\text{mi}}{\text{min}}} = 15 \text{ min}$$

Therefore, it will take 15 minutes to make the journey.

13. A: Taylor's total income is $\$20,000 + \$10,000 = \$30,000$. Fifteen percent of this is $\frac{15}{100} = \frac{3}{20}$. So:

$$\frac{3}{20} \times \$30,000 = \frac{\$90,000}{20}$$

$$\frac{\$9000}{2} = \$4500$$

14. B: Since the answer will be in cubic feet rather than inches, the first step is to convert from inches to feet for the dimensions of the box. There are 12 inches per foot, so the box is $24 \div 12 = 2$ feet wide, $18 \div 12 = 1.5$ feet deep, and $12 \div 12 = 1$ foot high. The volume is the product of these three together:

$$2 \times 1.5 \times 1 = 3 \text{ cubic feet}$$

15. D: Kristen bought four DVDs, which would cost a total of $4 \times 15 = \$60$. She spent a total of $100, so she spent $\$100 - \$60 = \$40$ on CDs. Since they cost $10 each, she must have purchased $40 \div 10 = 4$ CDs.

16. D: This problem can be solved by setting up a proportion involving the given information and the unknown value. The proportion is:

$$\frac{21 \text{ pages}}{4 \text{ nights}} = \frac{140 \text{ pages}}{x \text{ nights}}$$

Solving the proportion by cross-multiplying, the equation becomes $21x = 4 \times 140$, where $x = 26.67$. Since it is not an exact number of nights, the answer is rounded up to 27 nights. Twenty-six nights would not give Sarah enough time.

17. D: This problem can be solved by using unit conversion. The initial units are miles per minute. The final units need to be feet per second. Converting miles to feet uses the equivalence statement 1 mile equals 5,280 feet. Converting minutes to seconds uses the equivalence statement 1 minute equals 60 seconds. Setting up the ratios to convert the units is shown in the following equation:

$$\frac{72 \text{ mi}}{90 \text{ min}} \times \frac{1 \text{ min}}{60 \text{ s}} \times \frac{5280 \text{ ft}}{1 \text{ mi}} = 70.4 \frac{\text{ft}}{\text{s}}$$

The initial units cancel out, and the new units are left.

18. C: The sum total percentage of a pie chart must equal 100%. Since the CD sales take up less than half of the chart and more than a quarter (25%), it can be determined to be 40% overall. This can also be measured with a protractor. The angle of a circle is 360°. Since 25% of 360° would be 90° and 50% would be 180°, the angle percentage of CD sales falls in between; therefore, it would be Choice *C*.

19. B: Since $850 is the price *after* a 20% discount, $850 represents 80% of the original price. To determine the original price, set up a proportion with the ratio of the sale price (850) to original price (unknown) equal to the ratio of sale percentage (where x represents the unknown original price):

$$\frac{850}{x} = \frac{80}{100}$$

To solve a proportion, cross multiply the numerators and denominators and set the products equal to each other:

$$(850)(100) = (80)(x)$$

Multiplying each side results in the equation $85,000 = 80x$.

To solve for x, divide both sides by 80: $\frac{85,000}{80} = \frac{80x}{80}$, resulting in $x = 1062.5$. Remember that x represents the original price. Subtracting the sale price from the original price ($1062.50 - $850) indicates that Frank saved $212.50.

20. C: $40N$ would be 4000% of N. It's possible to check that each of the others is actually 40% of N.

21. C: An x-intercept is the point where the graph crosses the x-axis. At this point, the value of y is 0. To determine if an equation has an x-intercept of -2, substitute -2 for x, and calculate the value of y. If the value of -2 for x corresponds with a y-value of 0, then the equation has an x-intercept of -2. The only answer choice that produces this result is Choice *C*:

$$0 = (-2)^2 + 5(-2) + 6$$

22. C: The conditional frequency of a girl being potty-trained is calculated by dividing the number of potty-trained girls by the total number of girls: $34 \div 52 = 0.65$. To determine the conditional probability, multiply the conditional frequency by 100: $0.65 \times 100 = 65\%$.

23. A: A sample statistic indicates information about the data that was collected (in this case, the heights of those surveyed). A population parameter describes an aspect of the entire population (in this case, all twelve-year-old boys in the United States). A confidence interval would consist of a range of heights likely to include the actual population parameter. Measurement error relates to the validity of the data that was collected.

24. C: To ensure valid results, samples should be taken across the entire scope of the study. Since all states are not equally populated, representing each state proportionately would result in a more accurate statistic.

25. D: Exponential functions can be written in the form: $y = a \times b^x$. The equation for an exponential function can be written given the y-intercept (a) and the growth rate (b).

The y-intercept is the output (y) when the input (x) equals zero. It can be thought of as an "original value," or starting point. The value of b is the rate at which the original value increases ($b > 1$) or decreases ($b < 1$).

In this scenario, the y-intercept, a, would be \$1200, and the growth rate, b, would be 1.01 (100% of the original value combined with 1% interest, or $100\% + 1\% = 101\% = 1.01$).

26. C: 85% of a number means multiplying that number by 0.85. So, $0.85 \times 20 = \frac{85}{100} \times \frac{20}{1}$, which can be simplified to:

$$\frac{17}{20} \times \frac{20}{1} = 17$$

27. A: To find the fraction of the bill that the first three people pay, the fractions need to be added, which means finding the common denominator. The common denominator will be 60:

$$\frac{1}{5} + \frac{1}{4} + \frac{1}{3} = \frac{12}{60} + \frac{15}{60} + \frac{20}{60} = \frac{47}{60}$$

The remainder of the bill is:

$$1 - \frac{47}{60} = \frac{60}{60} - \frac{47}{60} = \frac{13}{60}$$

28. D: $9\frac{3}{10}$. To convert a decimal to a fraction, remember that any number to the left of the decimal point will be a whole number. Then, sense 0.3 goes to the tenths place, it can be placed over 10.

29. A: Convert the mixed fractions to improper fractions: $\frac{11}{3} - \frac{9}{5}$. Subtract using 15 as a common denominator and rewrite to get rid of the improper fraction:

$$\frac{11}{3} - \frac{9}{5} = \frac{55}{15} - \frac{27}{15} = \frac{28}{15} = 1\frac{13}{15}$$

30. A: Dividing by 98 can be approximated by dividing by 100, which would mean shifting the decimal point of the numerator to the left by 2. The result is 4.2 and rounds to 4.

31. B: First, separate out and add the whole numbers from the mixed fractions:

$$4\frac{1}{3} + 3\frac{3}{4}$$

$$4 + 3 + \frac{1}{3} + \frac{3}{4}$$

$$7 + \frac{1}{3} + \frac{3}{4}$$

Adding the fractions gives:

$$\frac{1}{3} + \frac{3}{4}$$

$$\frac{4}{12} + \frac{9}{12}$$

$$\frac{13}{12}$$

$$1 + \frac{1}{12}$$

Thus:

$$7 + \frac{1}{3} + \frac{3}{4} = 7 + 1 + \frac{1}{12} = 8\frac{1}{12}$$

32. C: The average is calculated by adding all six numbers, then dividing by 6. The first five numbers have a sum of 25. If the total divided by 6 is equal to 6, then the total itself must be 36. The sixth number must be $36 - 25 = 11$.

33. C: The domain consists of all possible inputs, or x-values. The scenario states that the function approximates the population between the years 1900 and 2000. It also states that $x = 0$ represents the year 1900. Therefore, the year 2000 would be represented by $x = 100$. Only inputs between 0 and 100 are relevant in this case.

34. C: Nothing is added to x and y since the center is 0 and 5^2 is 25. Choice A is not the correct answer because you do not subtract the radius from x and y. Choice B is not the correct answer because you must square the radius on the right side of the equation. Choice D is not the correct answer because you do not add the radius to x and y in the equation.

35. D: Subtract the center from the x and y values of the equation and square the radius on the right side of the equation. Choice A is not the correct answer because you need to square the radius of the equation. Choice B is not the correct answer because you do not square the centers of the equation. Choice C is not the correct answer because you need to subtract (not add) the centers of the equation.

36. C: Plug in the values for x and y to discover that the solution works, which is:

$$(-3)^2 + (-4)^2 = 25$$

Choices A and B are not the correct answers since the solution works. Choice D is not the correct answer because there is enough information to tell where the given point lies on the circle.

37. D: The correct answer of 20 °C can be found using the appropriate temperature conversion formula:

$$°C = (°F - 32) \times \frac{5}{9}$$

38. D: The slope is given by the change in y divided by the change in x. Specifically, it's:

$$slope = \frac{y_2 - y_1}{x_2 - x_1}$$

The first point is (-5, -3), and the second point is (0, -1). Work from left to right when identifying coordinates. Thus the point on the left is point 1 (-5,-3) and the point on the right is point 2 (0,-1).

Now we need to just plug those numbers into the equation:

$$slope = \frac{-1 - (-3)}{0 - (-5)}$$

It can be simplified to:

$$slope = \frac{-1 + 3}{0 + 5}$$

$$slope = \frac{2}{5}$$

39. B: The figure is composed of three sides of a square and a semicircle. The sides of the square are simply added: $8 + 8 + 8 = 24$ feet. The circumference of a circle is found by the equation $C = 2\pi r$. The radius is 4, so the circumference of the circle is 25.132 feet. Only half of the circle makes up the outer border of the figure (part of the perimeter) so half of 25.132 feet is 12.566 feet. Therefore, the total perimeter is: 24 ft + 12.566 ft = 36.566 ft. The other answer choices use the incorrect formula or fail to include all of the necessary sides.

40. A: The first step is to determine the unknown, which is in terms of the length, l.

The second step is to translate the problem into the equation using the perimeter of a rectangle:

$$P = 2l + 2w$$

The width is the length minus 2 centimeters. The resulting equation is:

$$2l + 2(l - 2) = 44$$

The equation can be solved as follows:

$2l + 2l - 4 = 44$	Apply the distributive property on the left side of the equation
$4l - 4 = 44$	Combine like terms on the left side of the equation
$4l = 48$	Add 4 to both sides of the equation
$l = 12$	Divide both sides of the equation by 4

The length of the rectangle is 12 centimeters. The width is the length minus 2 centimeters, which is 10 centimeters. Checking the answers for length and width forms the following equation:

$$44 = 2(12) + 2(10)$$

The equation can be solved using the order of operations to form a true statement: $44 = 44$.

41. B: $3x^2 - 3x + 11$. By distributing the implied one in front of the first set of parentheses and the -1 in front of the second set of parentheses, the parenthesis can be eliminated:

$$1(5x^2 - 3x + 4) - 1(2x^2 - 7)$$

$$5x^2 - 3x + 4 - 2x^2 + 7$$

Next, like terms (same variables with same exponents) are combined by adding the coefficients and keeping the variables and their powers the same:

$$5x^2 - 3x + 4 - 2x^2 + 7 = 3x^2 - 3x + 11$$

42. A: This question involves the percent formula. Since we're beginning with a percent, also known as a number over 100, we'll put 39 on the right side of the equation:

$$\frac{x}{164} = \frac{39}{100}$$

Now, multiply 164 and 39 to get 6,396, which then needs to be divided by 100.

$$6,396 \div 100 = 63.96$$

43. C: Kimberley worked 4.5 hours at the rate of \$10/h and 1 hour at the rate of \$12/h. The problem states that her pay is rounded to the nearest hour, so the 4.5 hours would round up to 5 hours at the rate of \$10/h.

$$(5h) \times \left(\frac{\$10}{h}\right) + (1h) \times \left(\frac{\$12}{h}\right)$$

$$\$50 + \$12 = \$62$$

44. B: The first step is to calculate the difference between the larger value and the smaller value:

$$378 - 252 = 126$$

To calculate this difference as a percentage of the original value, and thus calculate the percentage *increase*, 126 is divided by 252, then this result is multiplied by 100 to find the percentage: 50%, or Choice *B*.

45. C: In order to find the percentage by which the value of the car has been reduced, the current cash value should be subtracted from the initial value and then the difference divided by the initial value. The result should be multiplied by 100 to find the percentage decrease.

$$\frac{20,000 - 8,000}{20,000} = 0.6$$

$$(0.6) \times 100 = 60\%$$

46. A: This problem can be solved by simple multiplication and addition. Since the sale date is over six years apart, 6 can be multiplied by 12 for the number of months in a year, and then the remaining 4 months can be added.

$$(6 \times 12) + 4 = ?$$

$$72 + 4 = 76$$

47. D: This problem can be solved using basic arithmetic. Xavier starts with 20 apples, then gives his sister half, so 20 divided by 2.

$$\frac{20}{2} = 10$$

He then gives his neighbor 6, so 6 is subtracted from 10.

$$10 - 6 = 4$$

Lastly, he uses 3/4 of his apples to make an apple pie, so to find remaining apples, the first step is to subtract 3/4 from one and then multiply the difference by 4.

$$\left(1 - \frac{3}{4}\right) \times 4 = ?$$

$$\left(\frac{4}{4} - \frac{3}{4}\right) \times 4 = ?$$

$$\left(\frac{1}{4}\right) \times 4 = 1$$

48. B: To find the value of n, multiply $\frac{5}{2}$ by the reciprocal of $\frac{1}{3}$ and simplify:

$$n = \frac{5}{2} \div \frac{1}{3} = \frac{5}{2} \times \frac{3}{1} = \frac{15}{2} = 7.5$$

49. A: The total fraction taken up by green and red shirts will be:

$$\frac{1}{3} + \frac{2}{5} = \frac{5}{15} + \frac{6}{15} = \frac{11}{15}$$

The remaining fraction is:

$$1 - \frac{11}{15} = \frac{15}{15} - \frac{11}{15} = \frac{4}{15}$$

50. C: If she has used 1/3 of the paint, she has 2/3 remaining. $2\frac{1}{2}$ gallons are the same as $\frac{5}{2}$ gallons. The calculation is:

$$\frac{2}{3} \times \frac{5}{2} = \frac{5}{3} = 1\frac{2}{3} \; gallons$$

51. C: The volume of a cylinder is $\pi r^2 h$, and $\pi \times 6^2 \times 2$ is $72\,\pi$ cm³. Choice *A* is not the correct answer because that is only $6^2 \times \pi$. Choice *B* is not the correct answer because that is $2^2 \times 6 \times \pi$. Choice *D* is not the correct answer because that is $2^3 \times 6 \times \pi$.

52. B: This answer is correct because $3^2 + 4^2$ is $9 + 16$, which is 25. Taking the square root of 25 is 5. Choice *A* is not the correct answer because that is $3 + 4$. Choice *C* is not the correct answer because that is stopping at $3^2 + 4^2$ is $9 + 16$, which is 25. Choice *D* is not the correct answer because that is 3×4.

53. A: Setting up a proportion is the easiest way to represent this situation. The proportion becomes $\frac{20}{x} = \frac{40}{100}$, where cross-multiplication can be used to solve for x. Here, $40x = 2000$, so $x = 50$.

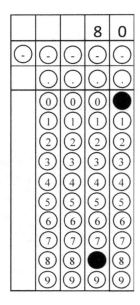

54. To solve the problem, a proportion is written consisting of ratios comparing distance and time. One way to set up the proportion is $\frac{3}{48} = \frac{5}{x}$ or $\left(\frac{distance}{time} = \frac{distance}{time}\right)$, where x represents the unknown value of time. To solve a proportion, the ratios are cross-multiplied:

$$(3)(x) = (5)(48)$$

$$3x = 240$$

The equation is solved by isolating the variable, or dividing by 3 on both sides, to produce $x = 80$.

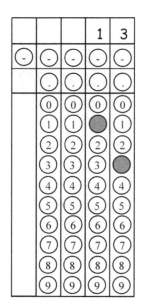

55. Perimeter is found by calculating the sum of all sides of the polygon. $9 + 9 + 9 + 8 + 8 + s = 56$, where s is the missing side length. Therefore, 43 plus the missing side length is equal to 56. The missing side length is 13 cm.

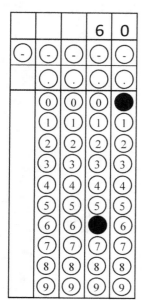

56. When x and y are complementary angles, the sine of x is equal to the cosine of y. The complementary angle of 30 is $90 - 30 = 60$ degrees. Therefore, the answer is 60 degrees.

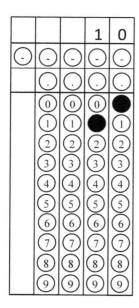

57. Start with the original equation: $x - 2xy + 2y$, then replace each instance of x with a 2, and each instance of y with a 3 to get:

$$2^2 - 2 \times 2 \times 3 + 2 \times 3^2$$

$$4 - 12 + 18 = 10$$

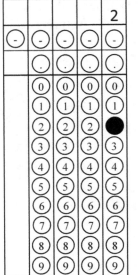

58. Add 3 to both sides to get $4x = 8$. Then divide both sides by 4 to get $x = 2$.

SAT Practice Test #3

Reading Test

Fiction

Questions 1–10 are based upon the following passage:

"Did you ever come across a protégé of his—one Hyde?" He asked.

"Hyde?" repeated Lanyon. "No. Never heard of him. Since my time."

That was the amount of information that the lawyer carried back with him to the great, dark bed on which he tossed to and fro until the small hours of the morning began to grow large. It was a night of little ease to his toiling mind, toiling in mere darkness and besieged by questions.

Six o'clock struck on the bells of the church that was so conveniently near to Mr. Utterson's dwelling, and still he was digging at the problem. Hitherto it had touched him on the intellectual side alone; but now his imagination also was engaged, or rather enslaved; and as he lay and tossed in the gross darkness of the night in the curtained room, Mr. Enfield's tale went by before his mind in a scroll of lighted pictures. He would be aware of the great field of lamps in a nocturnal city; then of the figure of a man walking swiftly; then of a child running from the doctor's; and then these met, and that human Juggernaut trod the child down and passed on regardless of her screams. Or else he would see a room in a rich house, where his friend lay asleep, dreaming and smiling at his dreams; and then the door of that room would be opened, the curtains of the bed plucked apart, the sleeper recalled, and, lo! There would stand by his side a figure to whom power was given, and even at that dead hour he must rise and do its bidding. The figure in these two phrases haunted the lawyer all night; and if at anytime he dozed over, it was but to see it glide more stealthily through sleeping houses, or move the more swiftly, and still the more smoothly, even to dizziness, through wider labyrinths of lamplighted city, and at every street corner crush a child and leave her screaming. And still the figure had no face by which he might know it; even in his dreams it had no face, or one that baffled him and melted before his eyes; and thus there it was that there sprung up and grew apace in the lawyer's mind a singularly strong, almost an inordinate, curiosity to behold the features of the real Mr. Hyde. If he could but once set eyes on him, he thought the mystery would lighten and perhaps roll altogether away, as was the habit of mysterious things when well examined. He might see a reason for his friend's strange preference or bondage, and even for the startling clauses of the will. And at least it would be a face worth seeing: the face of a man who was without bowels of mercy: a face which had but to show itself to raise up, in the mind of the unimpressionable Enfield, a spirit of enduring hatred.

From that time forward, Mr. Utterson began to haunt the door in the by-street of shops. In the morning before office hours, at noon when business was plenty and time scarce, at night under the face of the full city moon, by all lights and at all hours of solitude or concourse, the lawyer was to be found on his chosen post.

"If he be Mr. Hyde," he had thought, "I should be Mr. Seek."

Excerpt from *The Strange Case of Dr. Jekyll and Mr. Hyde* by Robert Louis Stevenson

1. What is the purpose of the use of repetition in the following passage?

It was a night of little ease to his toiling mind, toiling in mere darkness and besieged by questions.

a. It serves as a demonstration of the mental state of Mr. Lanyon.
b. It is reminiscent of the church bells that are mentioned in the story.
c. It mimics Mr. Utterson's ambivalence.
d. It emphasizes Mr. Utterson's anguish in failing to identify Hyde's whereabouts.

2. What is the setting of the story in this passage?
a. In the city
b. On the countryside
c. In a jail
d. In a mental health facility

3. What can one infer about the meaning of the word *Juggernaut* from the author's use of it in the passage?
a. It is an apparition that appears at daybreak.
b. It scares children.
c. It is associated with space travel.
d. Mr. Utterson finds it soothing.

4. What is the definition of the word *haunt* in the following passage?

From that time forward, Mr. Utterson began to haunt the door in the by-street of shops. In the morning before office hours, at noon when business was plenty and time scarce, at night under the face of the full city moon, by all lights and at all hours of solitude or concourse, the lawyer was to be found on his chosen post.

a. To levitate
b. To constantly visit
c. To terrorize
d. To daunt

5. The phrase *labyrinths of lamplighted city* contains an example of what?
a. Hyperbole
b. Simile
c. Juxtaposition
d. Alliteration

6. What can one reasonably conclude from the final comment of this passage?

"If he be Mr. Hyde," he had thought, "I should be Mr. Seek."

a. The speaker is considering a name change.
b. The speaker is experiencing an identity crisis.
c. The speaker has mistakenly been looking for the wrong person.
d. The speaker intends to continue to look for Hyde.

7. The author's attitude toward the main subject of this passage can be described as:
a. Intrigue
b. Elation
c. Animosity
d. Rigidity

8. According to the passage, what is Mr. Utterson struggling with as he tosses and turns in bed?
a. A murderer who is stalking Mr. Utterson since he moved to the city.
b. The mystery surrounding a dark figure and the terrible crimes he commits.
c. The cases he is involved in as a detective.
d. A chronic illness that is causing Mr. Utterson to hallucinate.

9. According to the passage, why did Mr. Utterson start to haunt the doors by the street shops?
a. He was looking for a long, lost love who he kept dreaming about.
b. He was recently homeless, and the street shops offered him food to eat when he was hungry.
c. He was looking for the dark, mysterious figure who he had been obsessing over in his sleep.
d. He was looking for a thief that would regularly steal out of stores.

10. What point of view is the passage written in?
a. First person
b. Second person
c. Third person limited
d. Third person omniscient

History/Social Studies

Questions 11–20 are based on the following passages:

Passage I

Lethal force, or deadly force, is defined as the physical means to cause death or serious harm to another individual. The law holds that lethal force is only accepted when you or another person are in immediate and unavoidable danger of death or severe bodily harm. For example, a person could be beating a weaker person in such a way that they are suffering severe enough trauma that could result in death or serious harm. This would be an instance where lethal force would be acceptable and possibly the only way to save that person from irrevocable damage.

Another example of when to use lethal force would be when someone enters your home with a deadly weapon. The intruder's presence and possession of the weapon indicate mal-intent and the ability to inflict death or severe injury to you and your loved ones. Again, lethal force can be used in this situation. Lethal force can also be applied to prevent the harm of another individual. If a woman is being brutally assaulted and is unable to fend off an attacker, lethal force can be used to defend her as a last-ditch effort. If she is in immediate jeopardy of rape, harm, and/or death, lethal force could be the only response that could effectively deter the assailant.

The key to understanding the concept of lethal force is the term *last resort*. Deadly force cannot be taken back; it should be used only to prevent severe harm or death. The law does distinguish whether the means of one's self-defense is fully warranted, or if the individual goes out of control in the process. If you continually attack the assailant after they are rendered incapacitated, this would be causing unnecessary harm, and the law can bring charges against you. Likewise, if you kill an attacker unnecessarily after defending yourself, you can be charged with murder. This would move lethal force beyond necessary defense, making it no longer a last resort but rather a use of excessive force.

Passage II

Assault is the unlawful attempt of one person to apply apprehension on another individual by an imminent threat or by initiating offensive contact. Assaults can vary, encompassing physical strikes, threatening body language, and even provocative language. In the case of the latter, even if a hand has not been laid, it is still considered an assault because of its threatening nature.

Let's look at an example: A homeowner is angered because his neighbor blows fallen leaves into his freshly mowed lawn. Irate, the homeowner gestures a fist to his fellow neighbor and threatens to bash his head in for littering on his lawn. The homeowner's physical motions and verbal threat heralds a physical threat against the other neighbor. These factors classify the homeowner's reaction as an assault. If the angry neighbor hits the threatening homeowner in retaliation, that would constitute an assault as well because he physically hit the homeowner.

Assault also centers on the involvement of weapons in a conflict. If someone fires a gun at another person, this could be interpreted as an assault unless the shooter acted in self-defense. If an individual drew a gun or a knife on someone with the intent to harm them, that would be considered assault. However, it's also considered an assault if someone simply aimed a weapon, loaded or not, at another person in a threatening manner.

11. What is the purpose of the second passage?
    a. To inform the reader about what assault is and how it is committed
    b. To inform the reader about how assault is a minor example of lethal force
    c. To disprove the previous passage concerning lethal force
    d. The author is recounting an incident in which they were assaulted

12. Which of the following situations, according to the passages, would not constitute an illegal use of lethal force?
    a. A disgruntled cashier yells obscenities at a customer.
    b. A thief is seen running away with stolen cash.
    c. A man is attacked in an alley by another man with a knife.
    d. A woman punches another woman in a bar.

13. Given the information in the passages, which of the following must be true about assault?
    a. Assault charges are more severe than unnecessary use of force charges.
    b. There are various forms of assault.
    c. Smaller, weaker people cannot commit assaults.
    d. Assault is justified only as a last resort.

14. Which of the following, if true, would most seriously undermine the explanation proposed by the author in Passage I in the third paragraph?
    a. An instance of lethal force in self-defense is not absolutely absolved from blame. The law considers the necessary use of force at the time it is committed.
    b. An individual who uses lethal force under necessary defense is in direct compliance of the law under most circumstances.
    c. Lethal force in self-defense should be forgiven in all cases for the peace of mind of the primary victim.
    d. The use of lethal force is not evaluated on the intent of the user but rather the severity of the primary attack that warranted self-defense.

15. Based on the passages, what can be inferred about the relationship between assault and lethal force?
    a. An act of lethal force always leads to a type of assault.
    b. An assault will result in someone using lethal force.
    c. An assault with deadly intent can lead to an individual using lethal force to preserve their well-being.
    d. If someone uses self-defense in a conflict, it is called deadly force; if actions or threats are intended, it is called assault.

16. Which of the following best describes the way the passages are structured?
    a. Both passages open by defining a legal concept and then continue to describe situations that further explain the concept.
    b. Both passages begin with situations, introduce accepted definitions, and then cite legal ramifications.
    c. Passage I presents a long definition while the Passage II begins by showing an example of assault.
    d. Both cite specific legal doctrines, then proceed to explain the rulings.

17. What can be inferred about the role of intent in lethal force and assault?
    a. Intent is irrelevant. The law does not take intent into account.
    b. Intent is vital for determining the lawfulness of using lethal force.
    c. Intent is very important for determining both lethal force and assault; intent is examined in both parties and helps determine the severity of the issue.
    d. The intent of the assailant is the main focus for determining legal ramifications; it is used to determine if the defender was justified in using force to respond.

18. The author uses the example in the second paragraph of Passage II in order to do what?
    a. To demonstrate two different types of assault by showing how each specifically relates to the other
    b. To demonstrate a single example of two different types of assault, then adding in the third type of assault in the example's conclusion
    c. To prove that the definition of lethal force is altered when the victim in question is a homeowner and his property is threatened
    d. To suggest that verbal assault can be an exaggerated crime by the law and does not necessarily lead to physical violence

19. As it is used in the second passage, the word *apprehension* most nearly means:
    a. Pain
    b. Exhaustion
    c. Fear
    d. Honor

20. One of the main purposes of the last paragraph in the first passage is to state:
    a. How assault is different when used in the home versus when it is used out in public.
    b. A specific example of lethal force so that the audience will know what it looks like.
    c. Why police officers defend those who use lethal force but do not defend those who use assault.
    d. The concept of lethal force as a last resort and the point at which it can cross a line from defense to manslaughter.

History/Social Studies

Questions 21–30 are based upon the following passage:

My Good Friends,—When I first imparted to the committee of the projected Institute my particular wish that on one of the evenings of my readings here the main body of my audience should be composed of working men and their families, I was animated by two desires; first, by the wish to have the great pleasure of meeting you face to face at this Christmas time, and accompany you myself through one of my little Christmas books; and second, by the wish to have an opportunity of stating publicly in your presence, and in the presence of the committee, my earnest hope that the Institute will, from the beginning, recognise one great principle—strong in reason and justice—which I believe to be essential to the very life of such an Institution. It is, that the working man shall, from the first unto the last, have a share in the management of an Institution which is designed for his benefit, and which calls itself by his name.

I have no fear here of being misunderstood—of being supposed to mean too much in this. If there ever was a time when any one class could of itself do much for its own good, and for the welfare of society—which I greatly doubt—that time is unquestionably past. It is in the fusion of different classes, without confusion; in the bringing together of employers and employed; in the creating of a better common understanding among those whose interests are identical, who depend upon each other, who are vitally essential to each other, and who never can be in unnatural antagonism without deplorable results, that one of the chief principles of a Mechanics' Institution should consist. In this world, a great deal of the bitterness among us arises from an imperfect understanding of one another. Erect in Birmingham a great

Educational Institution, properly educational; educational of the feelings as well as of the reason; to which all orders of Birmingham men contribute; in which all orders of Birmingham men meet; wherein all orders of Birmingham men are faithfully represented—and you will erect a Temple of Concord here which will be a model edifice to the whole of England.

Contemplating as I do the existence of the Artisans' Committee, which not long ago considered the establishment of the Institute so sensibly, and supported it so heartily, I earnestly entreat the gentlemen—earnest I know in the good work, and who are now among us—by all means to avoid the great shortcoming of similar institutions; and in asking the working man for his confidence, to set him the great example and give him theirs in return. You will judge for yourselves if I promise too much for the working man, when I say that he will stand by such an enterprise with the utmost of his patience, his perseverance, sense, and support; that I am sure he will need no charitable aid or condescending patronage; but will readily and cheerfully pay for the advantages which it confers; that he will prepare himself in individual cases where he feels that the adverse circumstances around him have rendered it necessary; in a word, that he will feel his responsibility like an honest man, and will most honestly and manfully discharge it. I now proceed to the pleasant task to which I assure you I have looked forward for a long time.

From Charles Dickens' speech in Birmingham in England on December 30, 1853 on behalf of the Birmingham and Midland Institute.

21. Which word is most closely synonymous with the word *patronage* as it appears in the following statement?

> ...that I am sure he will need no charitable aid or condescending patronage

a. Auspices
b. Aberration
c. Acerbic
d. Adulation

22. Which term is most closely aligned with the definition of the term *working man* as it is defined in the following passage?

> You will judge for yourselves if I promise too much for the working man, when I say that he will stand by such an enterprise with the utmost of his patience, his perseverance, sense, and support...

a. Plebeian
b. Viscount
c. Entrepreneur
d. Bourgeois

23. Which of the following statements most closely correlates with the definition of the term *working man* as it is defined in Question 22?
    a. A working man is not someone who works for institutions or corporations, but someone who is well-versed in the workings of the soul.
    b. A working man is someone who is probably not involved in social activities because the physical demand for work is too high.
    c. A working man is someone who works for wages among the middle class.
    d. The working man has historically taken to the field, to the factory, and now to the screen.

24. Based upon the contextual evidence provided in the passage above, what is the meaning of the term *enterprise* in the third paragraph?
    a. Company
    b. Courage
    c. Game
    d. Cause

25. The speaker addresses his audience as *My Good Friends.* What kind of credibility does this salutation give to the speaker?
    a. The speaker is an employer addressing his employees, so the salutation is a way for the boss to bridge the gap between himself and his employees.
    b. The speaker's salutation is one from an entertainer to his audience and uses the friendly language to connect to his audience before a serious speech.
    c. The salutation is used ironically to give a somber tone to the serious speech that follows.
    d. The speech is one from a politician to the public, so the salutation is used to grab the audience's attention.

26. According to the passage, what is the speaker's second desire for his time in front of the audience?
    a. To read a Christmas story
    b. For the working man to have a say in his institution, which is designed for his benefit
    c. To have an opportunity to stand in their presence
    d. For the life of the institution to be essential to the audience as a whole

27. The speaker's tone in the passage can be described as:
    a. Happy and gullible
    b. Lazy and entitled
    c. Confident and informed
    d. Angry and frustrated

28. One of the main purposes of the last paragraph is:
    a. To persuade the audience to support the Institute no matter what since it provided so much support to the working class.
    b. To market the speaker's new book while at the same time supporting the activities of the Institute.
    c. To inform the audience that the Institute is corrupt and will not help them out when the time comes to give them compensation.
    d. To provide credibility to the working man and share confidence in their ability to take on responsibilities if they are compensated appropriately.

29. According to the passage, what does the speaker wish to erect in Birmingham?
    a. An Educational Institution
    b. The Temple of Concord
    c. A Writing Workshop
    d. A VA Hospital

30. As it is used in the second paragraph, the word *antagonism* most nearly means:
    a. Conformity
    b. Opposition
    c. Affluence
    d. Scarcity

Science

Questions 31–40 are based upon the following passage:

Three years ago, I think there were not many bird-lovers in the United States who believed it possible to prevent the total extinction of both egrets from our fauna. All the known rookeries accessible to plume-hunters had been totally destroyed. Two years ago, the secret discovery of several small, hidden colonies prompted William Dutcher, President of the National Association of Audubon Societies, and Mr. T. Gilbert Pearson, Secretary, to attempt the protection of those colonies. With a fund contributed for the purpose, wardens were hired and duly commissioned. As previously stated, one of those wardens was shot dead in cold blood by a plume hunter. The task of guarding swamp rookeries from the attacks of money-hungry desperadoes to whom the accursed plumes were worth their weight in gold, is a very chancy proceeding. There is now one warden in Florida who says that "before they get my rookery they will first have to get me."

Thus far the protective work of the Audubon Association has been successful. Now there are twenty colonies, which contain all told, about 5,000 egrets and about 120,000 herons and ibises which are guarded by the Audubon wardens. One of the most important is on Bird Island, a mile out in Orange Lake, central Florida, and it is ably defended by Oscar E. Baynard. To-day, the plume hunters who do not dare to raid the guarded rookeries are trying to study out the lines of flight of the birds, to and from their feeding-grounds, and shoot them in transit. Their motto is—"Anything to beat the law, and get the plumes." It is there that the state of Florida should take part in the war.

The success of this campaign is attested by the fact that last year a number of egrets were seen in eastern Massachusetts—for the first time in many years. And so to-day the question is, can the wardens continue to hold the plume-hunters at bay?

Excerpt from Our Vanishing Wildlife by William T. Hornaday

31. The author's use of first-person pronouns in the following text does NOT have which of the following effects?

> Three years ago, I think there were not many bird-lovers in the United States who believed it possible to prevent the total extinction of both egrets from our fauna.

a. The phrase *I think* acts as a sort of hedging, where the author's tone is less direct and/or absolute.
b. It allows the reader to more easily connect with the author.
c. It encourages the reader to empathize with the egrets.
d. It distances the reader from the text by overemphasizing the story.

32. What purpose does the quote serve at the end of the first paragraph?
a. The quote shows proof of a hunter threatening one of the wardens.
b. The quote lightens the mood by illustrating the colloquial language of the region.
c. The quote provides an example of a warden protecting one of the colonies.
d. The quote provides much needed comic relief in the form of a joke.

33. What is the meaning of the word *rookeries* in the following text?

> To-day, the plume hunters who do not dare to raid the guarded rookeries are trying to study out the lines of flight of the birds, to and from their feeding-grounds, and shoot them in transit.

a. Houses in a slum area
b. A place where hunters gather to trade tools
c. A place where wardens go to trade stories
d. A colony of breeding birds

34. What is on Bird Island?
a. Hunters selling plumes
b. An important bird colony
c. Bird Island Battle between the hunters and the wardens
d. An important egret with unique plumes

35. What is the main purpose of the passage?
a. To persuade the audience to act in preservation of the bird colonies
b. To show the effect hunting egrets has had on the environment
c. To argue that the preservation of bird colonies has had a negative impact on the environment
d. To demonstrate the success of the protective work of the Audubon Association

36. According to the passage, why are hunters trying to study the lines of flight of the birds?
a. To study ornithology, one must know the lines of flight that birds take.
b. To help wardens preserve the lives of the birds
c. To have a better opportunity to hunt the birds
d. To build their homes under the lines of flight because they believe it brings good luck

37. A year before the passage was written, where were a number of egrets seen?
a. California
b. Florida
c. Eastern Massachusetts
d. Western Texas

38. As it is used in the first paragraph, the word *commissioned* most nearly means:
    a. Appointed
    b. Compelled
    c. Beguiled
    d. Fortified

39. What happened two years before the passage was written?
    a. The plume hunters didn't dare to raid the rookeries, as they are heavily guarded.
    b. Twenty colonies have emerged as thousands of egrets are protected and make their homes in safe havens.
    c. The plume hunters tried to shoot the birds in their line of flight.
    d. Several hidden colonies were found which prompted Dutcher and Pearson to protect them.

40. As it is used in the second paragraph, the phrase *in transit* most nearly means:
    a. On a journey or trip
    b. To give authority to
    c. On the way to the destination
    d. To make angry

Science

Questions 41–50 are based upon the following passage:

Insects as a whole are preeminently creatures of the land and the air. This is shown not only by the possession of wings by a vast majority of the class, but by the mode of breathing to which reference has already been made, a system of branching air-tubes carrying atmospheric air with its combustion-supporting oxygen to all the insect's tissues. The air gains access to these tubes through a number of paired air-holes or spiracles, arranged segmentally in series.

It is of great interest to find that, nevertheless, a number of insects spend much of their time under water. This is true of not a few in the perfect winged state, as for example aquatic beetles and water-bugs ('boatmen' and 'scorpions') which have some way of protecting their spiracles when submerged, and, possessing usually the power of flight, can pass on occasion from pond or stream to upper air. But it is advisable in connection with our present subject to dwell especially on some insects that remain continually under water till they are ready to undergo their final moult and attain the winged state, which they pass entirely in the air. The preparatory instars of such insects are aquatic; the adult instar is aerial. All may-flies, dragon-flies, and caddis-flies, many beetles and two-winged flies, and a few moths thus divide their life-story between the water and the air. For the present we confine attention to the Stone-flies, the May-flies, and the Dragon-flies, three well-known orders of insects respectively called by systematists the Plecoptera, the Ephemeroptera and the Odonata.

In the case of many insects that have aquatic larvae, the latter are provided with some arrangement for enabling them to reach atmospheric air through the surface-film of the water. But the larva of a stone-fly, a dragon-fly, or a may-fly is adapted more completely than these for aquatic life; it can, by means of gills of some kind, breathe the air dissolved in water.

This excerpt is from The Life-Story of Insects by Geo H. Carpenter

41. Which statement best details the central idea in this passage?
    a. It introduces certain insects that transition from water to air.
    b. It delves into entomology, especially where gills are concerned.
    c. It defines what constitutes as insects' breathing.
    d. It invites readers to have a hand in the preservation of insects.

42. Which definition most closely relates to the usage of the word *moult* in the passage?
    a. An adventure of sorts, especially underwater
    b. Mating act between two insects
    c. The act of shedding part or all of the outer shell
    d. Death of an organism that ends in a revival of life

43. What is the purpose of the first paragraph in relation to the second paragraph?
    a. The first paragraph serves as a cause, and the second paragraph serves as an effect.
    b. The first paragraph serves as a contrast to the second.
    c. The first paragraph is a description for the argument in the second paragraph.
    d. The first and second paragraphs are merely presented in a sequence.

44. What does the following sentence most nearly mean?

    The preparatory instars of such insects are aquatic; the adult instar is aerial.

    a. The volume of water is necessary to prep the insect for transition rather than the volume of the air.
    b. The abdomen of the insect is designed like a star in the water as well as the air.
    c. The stage of preparation in between molting is acted out in the water, while the last stage is in the air.
    d. These insects breathe first in the water through gills yet continue to use the same organs to breathe in the air.

45. Which of the statements reflect information that one could reasonably infer based on the author's tone?
    a. The author's tone is persuasive and attempts to call the audience to action.
    b. The author's tone is passionate due to excitement over the subject and personal narrative.
    c. The author's tone is informative and exhibits interest in the subject of the study.
    d. The author's tone is somber, depicting some anger at the state of insect larvae.

46. Which statement best describes stoneflies, mayflies, and dragonflies?
    a. They are creatures of the land and the air.
    b. They have a way of protecting their spiracles when submerged.
    c. Their larvae can breathe the air dissolved in water through gills of some kind.
    d. The preparatory instars of these insects are aerial.

47. According to the passage, what is true of "boatmen" and "scorpions"?
    a. They have no way of protecting their spiracles when submerged.
    b. They have some way of protecting their spiracles when submerged.
    c. They usually do not possess the power of flight.
    d. They remain continually under water till they are ready to undergo their final moult.

48. The last paragraph indicates that the author believes
    a. That the stonefly, dragonfly, and mayfly larvae are better prepared to live beneath the water because they have gills that allow them to do so.
    b. That the stonefly is different from the mayfly because the stonefly can breathe underwater and the mayfly can only breathe above water.
    c. That the dragonfly is a unique species in that its larvae lives mostly underwater for most of its young life.
    d. That the stonefly larvae can breathe only by reaching the surface film of the water.

49. According to the passage, why are insects as a whole preeminently creatures of the land and the air?
    a. Because insects are born on land but eventually end up adapting to life underwater for the rest of their adult lives.
    b. Because most insects have legs made for walking on land and tube-like structures on their bellies for skimming the water.
    c. Because a vast majority of insects have wings and also have the ability to breathe underwater.
    d. Because most insects have a propulsion method specifically designed for underwater use, but they can only breathe on land.

50. As it is used in the first paragraph, the word *preeminently* most nearly means:
    a. Unknowingly
    b. Above all
    c. Most truthfully
    d. Not importantly

Writing and Language Test

Aircraft Engineers

The knowledge of an aircraft engineer is acquired through years of education, and special licenses are required. Ideally, an individual will begin his or her preparation for the profession in high school by taking chemistry, physics, trigonometry, and calculus. Such curricula will aid in one's pursuit of a bachelor's degree in aircraft engineering, which requires several physical and life sciences, mathematics, and design courses.

(2) Some of universities provide internship or apprentice opportunities for the students enrolled in aircraft engineer programs. A bachelor's in aircraft engineering is commonly accompanied by a master's degree in advanced engineering or business administration. Such advanced degrees enable an individual to position himself or herself for executive, faculty, and/or research opportunities. (3) These advanced offices oftentimes require a Professional Engineering (PE) license which can be obtained through additional college courses, professional experience, and acceptable scores on the Fundamentals of Engineering (FE) and Professional Engineering (PE) standardized assessments.

Once the job begins, this line of work requires critical thinking, business skills, problem solving, and creativity. This level of (5) <u>expertise</u> (6) <u>allows</u> aircraft engineers to apply mathematical equations and scientific processes to aeronautical and aerospace issues or inventions. (8) <u>For example, aircraft engineers may test, design, and construct flying vessels such as airplanes, space shuttles, and missile weapons.</u> As a result, aircraft engineers are compensated with generous salaries. In fact, in May 2014, the lowest 10 percent of all American aircraft engineers earned less than $60,110 while the highest paid ten-percent of all American aircraft engineers earned $155,240. (9) <u>In May 2015, the United States Bureau of Labor Statistics (BLS) reported that the median annual salary of aircraft engineers was $107,830.</u> (10) <u>Conversely,</u> (11) <u>employment opportunities for aircraft engineers are projected to decrease by 2 percent by 2024.</u> This decrease may be the result of a decline in the manufacturing industry. Nevertheless, aircraft engineers who know how to utilize modeling and simulation programs, fluid dynamic software, and robotic engineering tools are projected to remain the most employable.

2015 Annual Salary of Aerospace Engineers

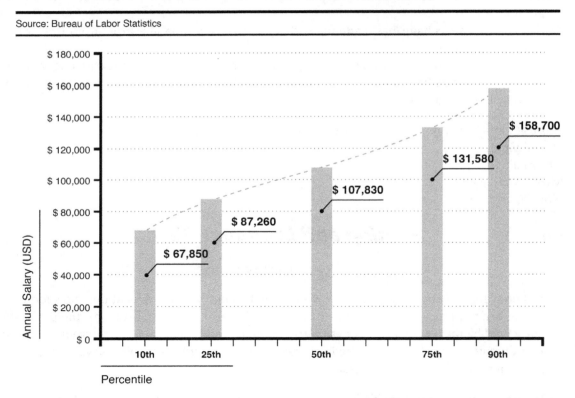

Source: Bureau of Labor Statistics

1. What type of text is utilized in the passage?
    a. Argumentative
    b. Narrative
    c. Biographical
    d. Informative

2. Which of the following would be the best choice for this sentence (reproduced below)?

(2) <u>Some of universities provide internship or apprentice opportunities</u> for the students enrolled in aircraft engineer programs.

a. NO CHANGE
b. Some of universities provided internship or apprentice opportunities
c. Some of universities provide internship or apprenticeship opportunities
d. Some universities provide internship or apprenticeship opportunities

3. Which of the following would be the best choice for this sentence (reproduced below)?

(3) <u>These advanced offices oftentimes require a Professional Engineering (PE) license which can be obtained through additional college courses, professional experience, and acceptable scores on the Fundamentals of Engineering (FE) and Professional Engineering (PE) standardized assessments.</u>

a. NO CHANGE
b. These advanced positions oftentimes require acceptable scores on the Fundamentals of Engineering (FE) and Professional Engineering (PE) standardized assessments in order to achieve a Professional Engineering (PE) license. Additional college courses and professional experience help.
c. These advanced offices oftentimes require acceptable scores on the Fundamentals of Engineering (FE) and Professional Engineering (PE) standardized assessments to gain the Professional Engineering (PE) license which can be obtained through additional college courses, professional experience.
d. These advanced positions oftentimes require a Professional Engineering (PE) license which is obtained by acceptable scores on the Fundamentals of Engineering (FE) and Professional Engineering (PE) standardized assessments. Further education and professional experience can help prepare for the assessments.

4. "The knowledge of an aircraft engineer is acquired through years of education." Which statement serves to support this claim?
a. Aircraft engineers are compensated with generous salaries.
b. Such advanced degrees enable an individual to position himself or herself for executive, faculty, or research opportunities.
c. Ideally, an individual will begin his or her preparation for the profession in high school by taking chemistry, physics, trigonometry, and calculus.
d. Aircraft engineers who know how to utilize modeling and simulation programs, fluid dynamic software, and robotic engineering tools will be the most employable.

5. What is the meaning of "expertise" in the marked sentence?
a. Care
b. Skill
c. Work
d. Composition

6. Which of the following would be the best choice for this sentence (reproduced below)?

This level of expertise (6) <u>allows</u> aircraft engineers to apply mathematical equation and scientific processes to aeronautical and aerospace issues or inventions.

a. NO CHANGE
b. Inhibits
c. Requires
d. Should

7. In the third paragraph, which of the following claims is supported?
a. This line of work requires critical thinking, business skills, problem solving, and creativity.
b. Aircraft engineers are compensated with generous salaries.
c. The knowledge of an aircraft engineer is acquired through years of education.
d. Those who work hard are rewarded accordingly.

8. Which of the following would be the best choice for this sentence (reproduced below)?

(8) <u>For example,</u> aircraft engineers may test, design, and construct flying vessels such as airplanes, space shuttles, and missile weapons.

a. NO CHANGE
b. Therefore,
c. However,
d. Furthermore,

9. Which of the following would be the best choice for this sentence (reproduced below)?

(9) <u>In May 2015, the United States Bureau of Labor Statistics (BLS) reported that the median annual salary of aircraft engineers was $107,830.</u>

a. NO CHANGE
b. May of 2015, the United States Bureau of Labor Statistics (BLS) reported that the median annual salary of aircraft engineers was $107,830.
c. In May of 2015 the United States Bureau of Labor Statistics (BLS) reported that the median annual salary of aircraft engineers was $107,830.
d. In May, 2015, the United States Bureau of Labor Statistics (BLS) reported that the median annual salary of aircraft engineers was $107,830.

10. Which of the following would be the best choice for this sentence (reproduced below)?

(10) <u>Conversely,</u> employment opportunities for aircraft engineers are projected to decrease by 2 percent by 2024.

a. NO CHANGE
b. Similarly,
c. In other words,
d. Accordingly,

11. Which of the following would be the best choice for this sentence (reproduced below)?

Conversely, (11) <u>employment opportunities for aircraft engineers are projected to decrease by 2 percent by 2024.</u>

a. NO CHANGE
b. Employment opportunities for aircraft engineers will be projected to decrease by 2 percent by 2024.
c. Employment opportunities for aircraft engineers is projected to decrease by 2 percent by 2024.
d. Employment opportunities for aircraft engineers was projected to decrease by 2 percent by 2024.

Attacks of September 11th

On September 11th 2001, a group of terrorists hijacked four American airplanes. The terrorists crashed the planes into the World Trade Center in New York City, the Pentagon in Washington D.C., and a field in Pennsylvania. Nearly 3,000 people died during the attacks, which propelled the United States into a "War on Terror".

About the Terrorists

Terrorists commonly use fear and violence to achieve political goals. The nineteen terrorists who orchestrated and implemented the attacks of September 11th were militants associated with al-Qaeda, an Islamic extremist group founded by Osama bin Laden, Abdullah Azzam, and others in the late 1980s. (13) <u>Bin Laden orchestrated the attacks as a response to what he felt was American injustice against Islam and hatred towards Muslims.</u> In his words, "Terrorism against America deserves to be praised."

Islam is the religion of Muslims, who live mainly in South and Southwest Asia and Sub-Saharan Africa. The majority of Muslims practice Islam peacefully. However, fractures in Islam have led to the growth of Islamic extremists who strictly oppose Western influences. They seek to institute stringent Islamic law and destroy those who (15) <u>violate</u> Islamic code.

In November 2002, bin Laden provided the explicit motives for the 9/11 terror attacks. According to this list, America's support of Israel, military presence in Saudi Arabia, and other anti-Muslim actions were the causes.

The Timeline of the Attacks

The morning of September 11 began like any other for most Americans. Then, at 8:45 a.m., a Boeing 767 plane crashed into the north tower of the World Trade Center in New York City. Hundreds were instantly killed. Others were trapped on higher floors. The (17) <u>crash was initially thought to be</u> a freak accident. When a second plane flew directly into the south tower eighteen minutes later, it was determined that America was under attack.

At 9:45 a.m., a third plane slammed into the Pentagon, America's military headquarters in Washington D.C. The jet fuel of this plane caused a major fire and partial building collapse that resulted in nearly 200 deaths. By 10:00 a.m., the south tower of the World Trade Center collapsed. Thirty minutes later, the north tower followed suit.

While this was happening, a fourth plane that departed from New Jersey, United Flight 93, was hijacked. The passengers learned of the attacks that occurred in New York and Washington D.C. and realized that they faced the same fate as the other planes that crashed. The passengers were determined to overpower the terrorists in an effort to prevent the deaths of additional innocent American citizens. Although the passengers were successful in (18) diverging the plane, it crashed in a western Pennsylvania field and killed everyone on board. The plane's final target remains uncertain, but many believe that United Flight 93 was heading for the White House.

Heroes and Rescuers

Close to 3,000 people died in the World Trade Center attacks. This figure includes 343 New York City firefighters and paramedics, 23 New York City police officers, and 37 Port Authority officers. Nevertheless, thousands of men and women in service worked valiantly to evacuate the buildings, save trapped workers, extinguish infernos, uncover victims trapped in fallen rubble, and tend to nearly 10,000 injured individuals.

About 300 rescue dogs played a major role in the after-attack salvages. Working twelve-hour shifts, the dogs scoured the rubble and alerted paramedics when they found signs of life. While doing so, the dogs served as a source of comfort and therapy for the rescue teams.

Initial Impacts on America

The attacks of September 11, 2001 resulted in the immediate suspension of all air travel. No flights could take off from or land on American soil. American airports and airspace closed to all national and international flights. Therefore, over five hundred flights had to turn back or be redirected to other countries. Canada alone received 226 flights and thousands of stranded passengers. Needless to say, as canceled flights are rescheduled, air travel became backed up and chaotic for quite some time.

At the time of the attacks, George W. Bush was the president of the United States. President Bush announced that "We will make no distinction between the terrorists who committed these acts and those who harbor them." The rate of hate crimes against American Muslims spiked, despite President Bush's call for the country to treat them with respect.

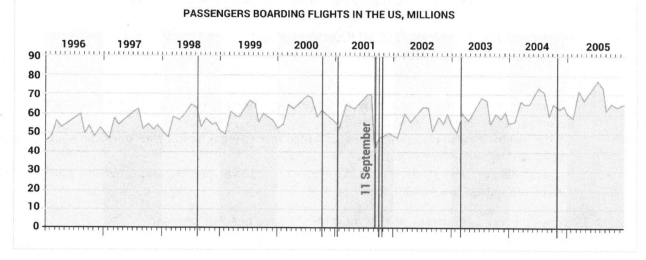

PASSENGERS BOARDING FLIGHTS IN THE US, MILLIONS

Additionally, relief funds were quickly arranged. The funds were used to support families of the victims, orphaned children, and those with major injuries. In this way, the tragic event brought the citizens together through acts of service towards those directly impacted by the attack.

Long-term Effects of the Attacks

Over the past fifteen years, the attacks of September 11th have transformed the United States' government, travel safety protocols, and international relations. Anti-terrorism legislation became a priority for many countries as law enforcement and intelligence agencies teamed up to find and defeat alleged terrorists.

Present George W. Bush announced a War on Terror. He (21) desired to bring bin Laden and al-Qaeda to justice and prevent future terrorist networks from gaining strength. The War in Afghanistan began in October of 2001 when the United States and British forces bombed al-Qaeda camps. (22) The Taliban, a group of fundamental Muslims who protected Osama bin Laden, was overthrown on December 9, 2001. However, the war continued in order to defeat insurgency campaigns in neighboring countries. Ten years later, the United State Navy SEALS killed Osama bin Laden in Pakistan. During 2014, the United States declared the end of its involvement in the War on Terror in Afghanistan.

Museums and memorials have since been erected to honor and remember the thousands of people who died during the September 11th attacks, including the brave rescue workers who gave their lives in the effort to help others.

12. How does the structure of the text help readers better understand the topic?
    a. By stating that anti-terrorism legislation was a priority for many countries, the reader can determine which laws were made and how they changed the life in the country.
    b. By placing the events in the order that they occurred, readers are better able to understand how the day unfolded.
    c. By using descriptive language, the readers are able to develop detailed images of the events that occurred during September 11, 2001.
    d. None of the above

13. Which of the following would be the best choice for this sentence (reproduced below)?

(13) <u>Bin Laden orchestrated the attacks as a response to what he felt was American injustice against Islam and hatred towards Muslims.</u>

a. NO CHANGE
b. Bin Laden orchestrated the attacks as a response to what he felt was American injustice against Islam, and hatred towards Muslims.
c. Bin Laden orchestrated the attacks, as a response to what he felt was American injustice against Islam and hatred towards Muslims.
d. Bin Laden orchestrated the attacks as responding to what he felt was American injustice against Islam and hatred towards Muslims.

14. How does the author express that most Muslims are peaceful people?
a. By describing the life of a Muslim after the attacks.
b. By including an anecdote about a Muslim friend.
c. By reciting details from religious texts.
d. By explicitly stating that fact.

15. What word could be used in exchange for "violate"?
a. Respect
b. Defile
c. Deny
d. Obey

16. What technique does the author use to highlight the impact of United Flight 93?
a. An image of the crash
b. An allusion to illustrate what may have occurred had the passengers not taken action
c. An anecdote about a specific passenger
d. A point of view consideration, where the author forces the reader to think about how he or she would have responded to such a situation

17. Which of the following would NOT be an appropriate replacement for the underlined portion of the sentence (reproduced below)?

The (17) <u>crash was initially thought to be </u>a freak accident.

a. First crash was thought to be
b. Initial crash was thought to be
c. Thought was that the initial crash
d. Initial thought was that the crash was

18. Which of the following would be the best choice for this sentence (reproduced below)?

Although the passengers were successful in (18) diverging the plane, it crashed in a western Pennsylvania field and killed everyone on board.

a. NO CHANGE
b. Diverting
c. Converging
d. Distracting

19. What statement is best supported by the graph included in this passage?
a. As canceled flights were rescheduled, air travel became backed up and chaotic for quite some time.
b. Over five hundred flights had to turn back or be redirected to other countries.
c. Canada alone received 226 flights and thousands of stranded passengers.
d. In the first few months following the attacks, there was a significant decrease in passengers boarding flights.

20. What is the purpose of the last paragraph?
a. It shows that beautiful art can be used to remember a past event.
b. It demonstrates that Americans will always remember the 9/11 attacks and the lives that were lost.
c. It explains how America fought back after the attacks.
d. It provides the author with an opportunity to explain how the location of the towers is used today.

21. Which of the following would be the best choice for this sentence (reproduced below)?

He (21) desired to bring bin Laden and al-Qaeda to justice and prevent future terrorist networks from gaining strength.

a. NO CHANGE
b. Perceived
c. Intended
d. Assimilated

22. Which of the following would be the best choice for this sentence (reproduced below)?

(22) The Taliban, a group of fundamental Muslims who protected Osama bin Laden, was overthrown on December 9, 2001. However, the war continued in order to defeat insurgency campaigns in neighboring countries.

a. NO CHANGE
b. The Taliban was overthrown on December 9, 2001. They were a group of fundamental Muslims who protected Osama bin Laden. However, the war continued in order to defeat insurgency campaigns in neighboring countries.
c. The Taliban, a group of fundamental Muslims who protected Osama bin Laden, on December 9, 2001 was overthrown. However, the war continued in order to defeat insurgency campaigns in neighboring countries.
d. Osama bin Laden's fundamental Muslims who protected him were called the Taliban and overthrown on December 9, 2001. Yet the war continued in order to defeat the insurgency campaigns in neighboring countries.

Fred Hampton

Fred Hampton desired to see lasting social change for African American people through nonviolent means and community recognition. (23) As a result, he became an African American activist during the American Civil Rights Movement and led the Chicago chapter of the Black Panther Party.

Hampton's Education

Hampton was born and raised in Maywood of Chicago, Illinois in 1948. (24) Gifted academically and a natural athlete, he became a stellar baseball player in high school. After graduating from Proviso East High School in 1966, he later went on to study law at Triton Junior College.

While studying at Triton, Hampton joined and became a leader of the National Association for the Advancement of Colored People (NAACP). (25) As a result of his leadership, the NAACP gained more than 500 members. Hampton worked relentlessly to acquire recreational facilities in the neighborhood and improve the educational resources provided to the impoverished black community of Maywood.

The Black Panthers

The Black Panther Party (BPP) was another activist group that formed around the same time as the NAACP. Hampton was quickly attracted to the Black Panther's approach to the fight for equal rights for African Americans. (26) Hampton eventually joined the chapter and relocated to downtown Chicago to be closer to its headquarters.

His (27) charismatic personality, organizational abilities, sheer determination, and rhetorical skills enabled him to quickly rise through the chapter's ranks. (28) Hampton soon became the leader of the Chicago chapter of the BPP where he organized rallies, taught political education classes, and established a free medical clinic. He also took part in the community police supervision project and played an instrumental role in the BPP breakfast program for impoverished African American children.

Hampton's greatest achievement as the leader of the BPP may be his fight against street gang violence in Chicago. In 1969, Hampton held a press conference where he made the gangs agree to a nonaggression pact known as the Rainbow Coalition. As a result of the pact, a multiracial alliance between blacks, Puerto Ricans, and poor youth was developed.

Assassination

As the Black Panther Party's popularity and influence grew, the Federal Bureau of Investigation (FBI) placed the group under constant surveillance. In an attempt to (30) neutralize the party, the FBI launched several harassment campaigns against the BPP, raided its headquarters in Chicago three times, and arrested over one hundred of the group's members. Hampton was shot during such a raid that occurred on the morning of December 4th, 1969.

In 1976, seven years after the event, it was revealed that William O'Neal, Hampton's trusted bodyguard, was an undercover FBI agent. (31) O'Neal provided the FBI with detailed floor plans of the BPP's headquarters, identifying the exact location of Hampton's bed. It was because of these floor plans that the police were able to target and kill Hampton.

The assassination of Hampton fueled outrage amongst the African American community. It was not until years after the assassination that the police admitted wrongdoing. The Chicago City Council now (32) commemorates December 4th as Fred Hampton Day.

23. Which of the following would be the best choice for this sentence (reproduced below)?

(23) As a result, he became an African American activist during the American Civil Rights Movement and led the Chicago chapter of the Black Panther Party.

a. NO CHANGE
b. As a result he became an African American activist
c. As a result: he became an African American activist
d. As a result of, he became an African American activist

24. What word could be used in place of the underlined description?

(24) Gifted academically and a natural athlete, he became a stellar baseball player in high school.

a. Vacuous
b. Energetic
c. Intelligent
d. Athletic

25. Which of the following statements, if true, would further validate the selected sentence?
a. Several of these new members went on to earn scholarships.
b. With this increase in numbers, Hampton was awarded a medal for his contribution to the NAACP.
c. This increase in membership was unprecedented in the NAACP's history.
d. The NAACP has been growing steadily every year.

26. How else could this sentence be re-structured while maintaining the context of the fourth paragraph?
    a. NO CHANGE
    b. Eventually, Hampton joined the chapter and relocated to downtown Chicago to be closer to its headquarters.
    c. Nevertheless, Hampton joined the chapter and relocated to downtown Chicago to be closer to its headquarters.
    d. Hampton then joined the chapter and relocated to downtown Chicago to be closer to its headquarters

27. What word is synonymous with the underlined description?
    a. Egotistical
    b. Obnoxious
    c. Chauvinistic
    d. Charming

28. Which of the following would be the best choice for this sentence (reproduced below)?

    (28) Hampton soon became the leader of the Chicago chapter of the BPP where he organized rallies, taught political education classes, and established a free medical clinic.

    a. NO CHANGE
    b. As the leader of the BPP, Hampton: organized rallies, taught political education classes, and established a free medical clinic.
    c. As the leader of the BPP, Hampton; organized rallies, taught political education classes, and established a free medical clinic.
    d. As the leader of the BPP, Hampton—organized rallies, taught political education classes, and established a medical free clinic.

29. The author develops the idea that Frank Hampton should not have been killed at the hands of the police. Which could best be used to support that claim?
    a. The manner in which the police raided the BPP headquarters.
    b. The eventual admission from the police that they were wrong in killing Hampton.
    c. The description of previous police raids that resulted in the arrest of hundreds BPP members.
    d. All of the above.

30. Which of the following would be the best choice for this sentence (reproduced below)?

    In an attempt to (30) neutralize the party, the FBI launched several harassment campaigns against the BPP, raided its headquarters in Chicago three times, and arrested over one hundred of the group's members.

    a. NO CHANGE
    b. Accommodate
    c. Assuage
    d. Praise

31. Which of the following would be the best choice for this sentence (reproduced below)?

(31) O'Neal provided the FBI with detailed floor plans of the BPP's headquarters, identifying the exact location of Hampton's bed.

a. NO CHANGE
b. O'Neal provided the FBI with detailed floor plans of the BPP's headquarters, which identified the exact location of Hampton's bed.
c. O'Neal provided the FBI with detailed floor plans and Hampton's bed.
d. O'Neal identified the exact location of Hampton's bed that provided the FBI with detailed floor plans of the BPP's headquarters.

32. What word could be used in place of the underlined word?
a. Disregards
b. Memorializes
c. Communicates
d. Denies

33. How would the author likely describe the FBI during the events of the passage?
a. Corrupt
b. Confused
c. Well-intended
d. Prejudiced

Here Comes the Flood!

A flood occurs when an area of land that is normally dry becomes submerged with water. Floods have affected Earth since the beginning of time and are caused by many different factors. (36) Flooding can occur slowly or within seconds and can submerge small regions or extend over vast areas of land. Their impact on society and the environment can be harmful or helpful.

What Causes Flooding?

Floods may be caused by natural phenomenon, induced by the activities of humans and other animals, or the failure of an infrastructure. Areas located near bodies of water are prone to flooding as are low-lying regions.

Global warming is the result of air pollution that prevents the sun's radiation from being emitted back into space. Instead, the radiation is trapped in Earth and results in global warming. The warming of the Earth has resulted in climate changes. As a result, floods have been occurring with increasing regularity. Some claim that the increased temperatures on Earth may cause the icebergs to melt. They fear that the melting of icebergs will cause the (37) <u>oceans levels</u> to rise and flood coastal regions.

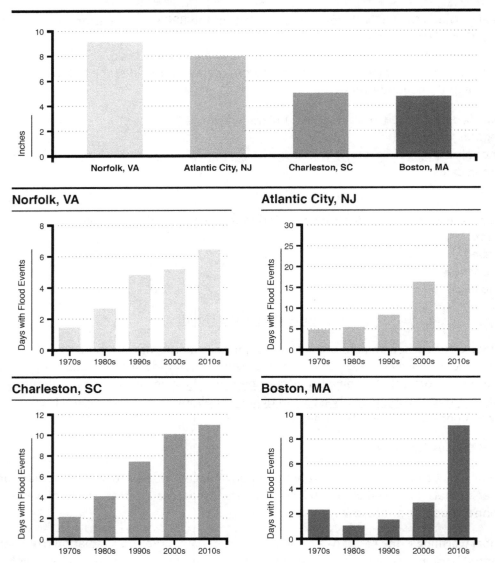

Local Sea Level Rise and Tidal Flooding, 1970-2012

Most commonly, flooding is caused by excessive rain. The ground is not able to absorb all the water produced by a sudden heavy rainfall or rainfall that occurs over a prolonged period of time. Such rainfall may cause the water in rivers and other bodies of water to overflow. The excess water can cause dams to break. Such events can cause flooding of the surrounding riverbanks or coastal regions.

Flash flooding can occur without warning and without rainfall. Flash floods may be caused by a river being blocked by a glacier, avalanche, landslide, logjam, a beaver's obstruction, construction, or dam. Water builds behind such a blockage. Eventually, the mass and force of the built-up water become so extreme that it causes the obstruction to break. Thus, enormous amounts of water rush out towards the surrounding areas.

Areal or urban flooding occurs because the land has become hardened. The hardening of land may result from urbanization or drought. Either way, the hardened land prevents water from seeping into the ground. Instead, the water resides on top of the land.

Finally, flooding may result after severe hurricanes, tsunamis, or tropical cyclones. Local defenses and infrastructures are no matches for the tidal surges and waves caused by these natural phenomena. Such events are bound to result in the flooding of nearby coastal regions or estuaries.

(35) A Floods After-Effects

Flooding can result in severe devastation of nearby areas. Flash floods and tsunamis can result in sweeping waters that travel at destructive speeds. Fast-moving water has the power to demolish all obstacles in its path such as homes, trees, bridges, and buildings. Animals, plants, and humans may all lose their lives during a flood.

Floods can also cause pollution and infection. Sewage may seep from drains or septic tanks and contaminate drinking water or surrounding lands. Similarly, toxins, fuels, debris from annihilated buildings, and other hazardous materials can leave water unusable for consumption. (38) <u>As the water begins to drain, mold may begin to grow.</u> As a result, residents of flooded areas may be left without power, drinkable water, or be exposed to toxins and other diseases.

(39) <u>Although often associated with devastation, not all flooding results</u> in adverse circumstances. For thousands of years, peoples have inhabited floodplains of rivers. (41) <u>Examples include the Mississippi Valley of the United States, the Nile River in Egypt, and the Tigris River of the Middle East</u>. The flooding of such rivers (42) <u>caused</u> nutrient-rich silts to be deposited on the floodplains. Thus, after the floods recede, an extremely fertile soil is left behind. This soil is conducive to the agriculture of bountiful crops and has sustained the diets of humans for millennium.

Proactive Measures Against Flooding

Technologies now allow scientists to predict where and when flooding is likely to occur. Such technologies can also be used (43) <u>to project</u> the severity of an anticipated flood. In this way, local inhabitants can be warned and take preventative measures such as boarding up their homes, gathering necessary provisions, and moving themselves and possessions to higher grounds.

The (44) <u>picturesque</u> views of coastal regions and rivers have long enticed people to build near such locations. Due to the costs associated with the repairs needed after the flooding of such residencies, many governments now require inhabitants of flood-prone areas to purchase flood insurance and build flood-resistant structures. Pictures of all items within a building or home should be taken so that proper reimbursement for losses can be made in the event that a flood does occur.

Staying Safe During a Flood

If a forecasted flood does occur, then people should retreat to higher ground such as a mountain or roof. Flooded waters may be contaminated, contain hidden debris, or travel at high speeds. Therefore, people should not attempt to walk or drive through a flooded area. To prevent electrocution, electrical outlets and downed power lines need to be avoided.

The Flood Dries Up

Regardless of the type or cause of a flood, floods can result in detrimental alterations to nearby lands and serious injuries to nearby inhabitants. By understanding flood cycles, civilizations can learn to take advantage of flood seasons. By taking the proper precautionary measures, people can stay safe when floods occur. Thus, proper knowledge can lead to safety and prosperity during such an adverse natural phenomenon.

34. What information from the graphs could be used to support the claims found in the third paragraph?
    a. Between 1970–1980, Boston experienced an increase in the number of days with flood events.
    b. Between 1970–1980, Atlantic City, New Jersey did not experience an increase in the number of days with flood events.
    c. Since 1970, the number of days with floods has decreased in major coastal cities across America.
    d. Since 1970, sea levels have risen along the East Coast.

35. One of the headings is entitled "A Floods After-Effects." How should this heading be rewritten?
    a. A Flood's After-Effect
    b. A Flood's After-Effects
    c. A Floods After-Affect
    d. A Flood's After-Affects

36. Which of the following revisions can be made to the sentence (reproduced below) that will still maintain the original meaning while making the sentence more concise?

(36) Flooding can occur slowly or within seconds and can submerge small regions or extend over vast areas of land.

    a. NO CHANGE
    b. Flooding can either be slow or occur within seconds. It doesn't take long to submerge small regions or extend vast areas of land.
    c. Flooding occurs slowly or rapidly submerging vast areas of land.
    d. Vast areas of land can be flooded slowly or within seconds.

37. Which of the following would be the best choice for this sentence (reproduced below)?

They fear that the melting of icebergs will cause the (37) oceans levels to rise and flood coastal regions.

a. NO CHANGE
b. ocean levels
c. ocean's levels
d. levels of the oceans

38. Which choice best maintains the pattern of the first sentence of the paragraph?
a. NO CHANGE
b. As the rain subsides and the water begins to drain, mold may begin to grow.
c. Mold may begin to grow as the water begins to drain.
d. The water will begin to drain and mold will begin to grow.

39. Which of the following would be the best choice for this sentence (reproduced below)?

(39) Although often associated with devastation, not all flooding results in adverse circumstances. For thousands of years, peoples have inhabited floodplains of rivers.

a. NO CHANGE
b. Although often associated with devastation not all flooding results
c. Although often associated with devastation. Not all flooding results
d. While often associated with devastation, not all flooding results

40. What is the author's intent of the final paragraph?
a. To explain that all bad occurrences eventually come to an end.
b. To summarize the key points within the passage.
c. To explain that, with time, all flooded lands will eventually dry.
d. To relay a final key point about floods.

41. The author is considering deleting this sentence (reproduced below) from the tenth paragraph. Should the sentence be kept or deleted?

(41) Examples include the Mississippi Valley of the United States, the Nile River in Egypt, and the Tigris River of the Middle East.

a. Kept, because it provides examples of floodplains that have been successfully inhabited by civilizations.
b. Kept, because it provides an example of how floods can be beneficial.
c. Deleted, because it blurs the paragraph's focus on the benefits of floods.
d. Deleted, because it distracts from the overall meaning of the paragraph.

42. Which of the following would be the best choice for this sentence (reproduced below)?

The flooding of such rivers (42) <u>caused</u> nutrient-rich silts to be deposited on the floodplains.

a. NO CHANGE
b. Cause
c. Causing
d. Causes

43. Which of the following would be the best choice for this sentence (reproduced below)?

Such technologies can also be used (43) <u>to project</u> the severity of an anticipated flood.

a. NO CHANGE
b. Projecting
c. Project
d. Projected

44. Which term could best replace the underlined word (reproduced below)?

The (44) <u>picturesque</u> views of coastal regions and rivers have long enticed people to build near such locations.

a. Colorful
b. Drab
c. Scenic
d. Candid

Math Test

1. If $6t + 4 = 16$, what is t?
   a. 1
   b. 2
   c. 3
   d. 4

2. The variable y is directly proportional to x. If $y = 3$ when $x = 5$, then what is y when $x = 20$?
   a. 10
   b. 12
   c. 14
   d. 16

3. A line passes through the point (1, 2) and crosses the y-axis at $y = 1$. Which of the following is an equation for this line?
   a. $y = 2x$
   b. $y = x + 1$
   c. $x + y = 1$
   d. $y = \frac{x}{2} - 2$

4. There are $4x + 1$ treats in each party favor bag. If a total of $60x + 15$ treats are distributed, how many bags are given out?
  a. 15
  b. 16
  c. 20
  d. 22

5. Apples cost $2 each, while bananas cost $3 each. Maria purchased 10 fruits in total and spent $22. How many apples did she buy?
  a. 5
  b. 6
  c. 7
  d. 8

6. What are the polynomial roots of $x^2 + x - 2$?
  a. 1 and -2
  b. -1 and 2
  c. 2 and -2
  d. 9 and 13

7. What is the y-intercept of $y = x^{\frac{5}{3}} + (x - 3)(x + 1)$?
  a. 3.5
  b. 7.6
  c. -3
  d. -15.1

8. $x^4 - 16$ can be simplified to which of the following?
  a. $(x^2 - 4)(x^2 + 4)$
  b. $(x^2 + 4)(x^2 + 4)$
  c. $(x^2 - 4)(x^2 - 4)$
  d. $(x^2 - 2)(x^2 + 4)$

9. $(4x^2y^4)^{\frac{3}{2}}$ can be simplified to which of the following?
  a. $8x^3y^6$
  b. $4x^{\frac{5}{2}}y$
  c. $4xy$
  d. $32x^{\frac{7}{2}}y^{\frac{11}{2}}$

10. If $\sqrt{1 + x} = 4$, what is x?
  a. 10
  b. 15
  c. 20
  d. 25

11. Suppose $\frac{x+2}{x} = 2$. What is x?

    a. -1

    b. 0

    c. 2

    d. 4

12. A ball is thrown from the top of a high hill, so that the height of the ball as a function of time is $h(t) = -16t^2 + 4t + 6$, in feet. What is the maximum height of the ball in feet?

    a. 6

    b. 6.25

    c. 6.5

    d. 6.75

13. A rectangle has a length that is 5 feet longer than three times its width. If the perimeter is 90 feet, what is the length in feet?

    a. 10

    b. 20

    c. 25

    d. 35

14. Five students take a test. The scores of the first four students are 80, 85, 75, and 60. If the median score is 80, which of the following could NOT be the score of the fifth student?

    a. 60

    b. 80

    c. 85

    d. 100

15. In an office, there are 50 workers. A total of 60% of the workers are women, and the chances of a woman wearing a skirt is 50%. If no men wear skirts, how many workers are wearing skirts?

    a. 12

    b. 15

    c. 16

    d. 20

16. Ten students take a test. Five students get a 50. Four students get a 70. If the average score is 55, what was the last student's score?

    a. 20

    b. 40

    c. 50

    d. 60

17. A company invests \$50,000 in a building where they can produce saws. If the cost of producing one saw is \$40, then which function expresses the amount of money the company pays? The variable y is the money paid and x is the number of saws produced.

    a. $y = 50,000x + 40$

    b. $y + 40 = x - 50,000$

    c. $y = 40x - 50,000$

    d. $y = 40x + 50,000$

18. A six-sided die is rolled. What is the probability that the roll is 1 or 2?

  a. $\frac{1}{6}$

  b. $\frac{1}{4}$

  c. $\frac{1}{3}$

  d. $\frac{1}{2}$

19. A line passes through the origin and through the point (-3, 4). What is the slope of the line?

  a. $-\frac{4}{3}$

  b. $-\frac{3}{4}$

  c. $\frac{4}{3}$

  d. $\frac{3}{4}$

20. A pair of dice is thrown, and the sum of the two scores is calculated. What's the expected value of the roll?

  a. 5

  b. 6

  c. 7

  d. 8

21.

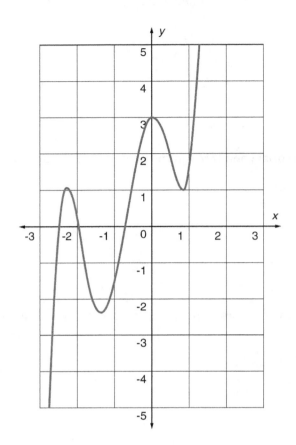

Which of the following functions represents the graph above?

    a. $y = x^5 + 3.5x^4 - 6.5x^2 + 0.5x + 3$
    b. $y = x^5 - 3.5x^4 + 6.5x^2 - 0.5x - 3$
    c. $y = 5x^4 - 6.5x^2 + 0.5x + 3$
    d. $y = -5x^4 - 6.5x^2 + 0.5x + 3$

22. Katie works at a clothing company and sold 192 shirts over the weekend. One third of the shirts that were sold were patterned, and the rest were solid. Which mathematical expression would calculate the number of solid shirts Katie sold over the weekend?

    a. $192 \times \frac{1}{3}$

    b. $192 \div \frac{1}{3}$

    c. $192 \times (1 - \frac{1}{3})$

    d. $192 \div 3$

23. Which measure for the center of a small sample set is most affected by outliers?

    a. Mean
    b. Median
    c. Mode
    d. None of the above

24. Given the value of a stock at monthly intervals, which graph should be used to best represent the trend of the stock?
    a. Box plot
    b. Line plot
    c. Line graph
    d. Circle graph

25. What is the probability of randomly picking the winner and runner-up from a race of 4 horses and distinguishing which is the winner?
    a. $\frac{1}{4}$

    b. $\frac{1}{2}$

    c. $\frac{1}{16}$

    d. $\frac{1}{12}$

26. What is the next number in the following series: $1, 3, 6, 10, 15, 21, \dots$?
    a. 26
    b. 27
    c. 28
    d. 29

27. A shipping box has a length of 8 inches, a width of 14 inches, and a height of 4 inches. If all three dimensions are doubled, what is the relationship between the volume of the new box and the volume of the original box?
    a. The volume of the new box is double the volume of the original box.
    b. The volume of the new box is four times as large as the volume of the original box.
    c. The volume of the new box is six times as large as the volume of the original box.
    d. The volume of the new box is eight times as large as the volume of the original box.

28. What is the product of the following expression?

$$(3 + 2i)(5 - 4i)$$

    a. $23 - 2i$
    b. $15 - 8i$
    c. $15 - 8i^2$
    d. $15 - 10i$

29.

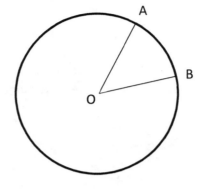

The length of arc $AB = 3\pi$ cm. The length of $\overline{OA} = 12$ cm. What is the degree measure of $\angle AOB$?
- a. 30 degrees
- b. 40 degrees
- c. 45 degrees
- d. 55 degrees

30.

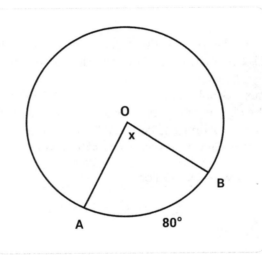

The area of circle O is 49π m. What is the area of the sector formed by $\angle AOB$?
- a. 80π m
- b. 10.9π m
- c. 4.9π m
- d. 10π m

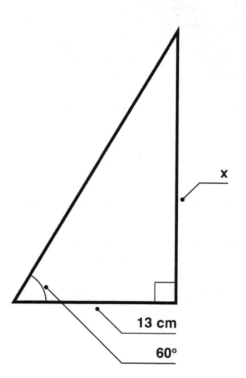
31. The triangle shown below is a right triangle. What's the value of x?

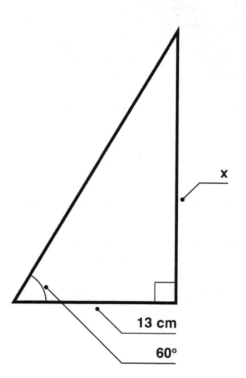

13 cm

60°

    a. $x = 1.73$
    b. $x = 0.57$
    c. $x = 13$
    d. $x = 22.52$

32. A ball is drawn at random from a ball pit containing 8 red balls, 7 yellow balls, 6 green balls, and 5 purple balls. What's the probability that the ball drawn is yellow?
    a. $\frac{1}{26}$

    b. $\frac{19}{26}$

    c. $\frac{7}{26}$

    d. 1

33. What's the probability of rolling a 6 at least once in two rolls of a die?
    a. $\frac{1}{3}$

    b. $\frac{1}{36}$

    c. $\frac{1}{6}$

    d. $\frac{11}{36}$

34. For a group of 20 men, the median weight is 180 pounds and the range is 30 pounds. If each man gains 10 pounds, which of the following would be true?
    a. The median weight will increase, and the range will remain the same.
    b. The median weight and range will both remain the same.
    c. The median weight will stay the same, and the range will increase.
    d. The median weight and range will both increase.

35. If the ordered pair $(-3, -4)$ is reflected over the x-axis, what's the new ordered pair?
    a. $(-3, -4)$
    b. $(3, -4)$
    c. $(3, 4)$
    d. $(-3, 4)$

36. If the volume of a sphere is 288π cubic meters, what are the radius and surface area of the same sphere?
    a. Radius 6 meters and surface area 144π square meters
    b. Radius 36 meters and surface area 144π square meter
    c. Radius 6 meters and surface area 12π square meters
    d. Radius 36 meters and surface area 12π square meters

37. Which four-sided shape is always a rectangle?
    a. Rhombus
    b. Square
    c. Parallelogram
    d. Quadrilateral

38. Using trigonometric ratios for a right angle, what is the value of the angle whose opposite side is equal to 25 centimeters and whose hypotenuse is equal to 50 centimeters?
    a. 15°
    b. 30°
    c. 45°
    d. 90°

39. Using trigonometric ratios for a right angle, what is the value of the closest angle whose adjacent side is equal to 7.071 centimeters and whose hypotenuse is equal to 10 centimeters?
    a. 15°
    b. 30°
    c. 45°
    d. 90°

40. Using trigonometric ratios, what is the value of an angle whose opposite side is equal to 1 inch and whose adjacent side is equal to the square root of 3 inches?
    a. 15°
    b. 30°
    c. 45°
    d. 90°

41. What is the function that forms an equivalent graph to $y = \cos(x)$?
   a. $y = \tan(x)$
   b. $y = \csc(x)$
   c. $y = \sin\left(x + \dfrac{\pi}{2}\right)$
   d. $y = \sin\left(x - \dfrac{\pi}{2}\right)$

42. A solution needs 5 mL of saline for every 8 mL of medicine given. How much saline is needed for 45 mL of medicine?
   a. $\dfrac{225}{8}$ mL
   b. 72 mL
   c. 28 mL
   d. $\dfrac{45}{8}$ mL

43. What's the midpoint of a line segment with endpoints $(-1, 2)$ and $(3, -6)$?
   a. $(1, 2)$
   b. $(1, 0)$
   c. $(-1, 2)$
   d. $(1, -2)$

44. A sample data set contains the following values: 1, 3, 5, 7. What's the standard deviation of the set?
   a. 2.58
   b. 4
   c. 6.23
   d. 1.1

No Calculator Questions

45. An equilateral triangle has a perimeter of 18 feet. If a square whose sides have the same length as one side of the triangle is built, what will be the area of the square?
   a. 6 square feet
   b. 36 square feet
   c. 256 square feet
   d. 1000 square feet

46. What is the volume of a sphere, in terms of π, with a radius of 3 inches?
   a. $36\,\pi$ in^3
   b. $27\,\pi$ in^3
   c. $9\,\pi$ in^3
   d. $72\,\pi$ in^3

47. What is the length of the other leg of a right triangle with a hypotenuse of 10 inches and a leg of 8 inches?
   a. 6 in
   b. 18 in
   c. 80 in
   d. 13 in

48. A pizzeria owner regularly creates jumbo pizzas, each with a radius of 9 inches. She is mathematically inclined, and wants to know the area of the pizza to purchase the correct boxes and know how much she is feeding her customers. What is the area of the circle, in terms of π, with a radius of 9 inches?
    a. 81π in^2
    b. 18π in^2
    c. 90π in^2
    d. 9π in^2

49. How will the following expression be written in standard form?

$$(1 \times 10^4) + (3 \times 10^3) + (7 \times 10^1) + (8 \times 10^0)$$

    a. 137
    b. 13,078
    c. 1,378
    d. 8,731

50. What is the simplified form of the expression $tan\theta \; cos\theta$?
    a. $sin\theta$
    b. 1
    c. $csc\theta$
    d. $\frac{1}{sec\theta}$

51. What is the value of the sum of $\frac{1}{3}$ and $\frac{2}{5}$?
    a. $\frac{3}{8}$
    b. $\frac{11}{15}$
    c. $\frac{11}{30}$
    d. $\frac{4}{5}$

52. If the cosine of $30° = x$, the sine of what angle also equals x?
    a. 30°
    b. 60°
    c. 90°
    d. 120°

53. If the tangent of $45° = x$, the sine of what angle also equals x?
    a. 30°
    b. 60°
    c. 90°
    d. 120°

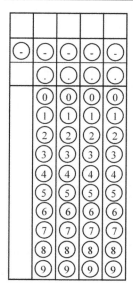

54. If $\overline{AE} = 4$, $\overline{AB} = 5$, and $\overline{AD} = 5$, what is the length of $\overline{AC}$?

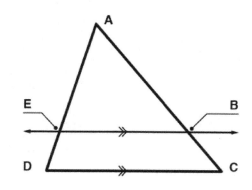

55. What is the decimal value of $\frac{3}{25}$?

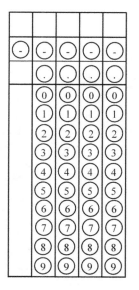

56. 6 is 30% of what number?

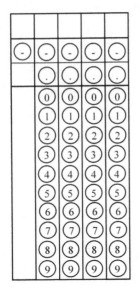

57. What is the value of the following expression?

$$\sqrt{8^2 + 6^2}$$

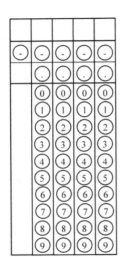

58. What is the measurement of angle f in the following picture? Assume the lines are parallel.

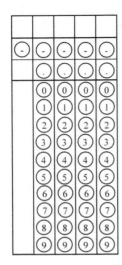

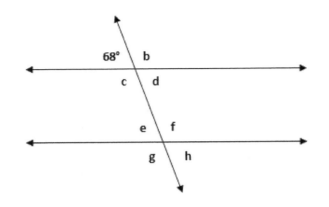

Essay Prompt

As you read the essay below, consider how the author uses these following things:

- Evidence to support claims, like facts or examples
- Reasoning to develop ideas and connect claims to evidence
- Stylistic elements, such as word choice or appeals to emotion, to express ideas

Dana Gioia argues in his article that poetry is dying, now little more than a limited art form confined to academic and college settings. Of course poetry remains healthy in the academic setting, but the idea of poetry being limited to this academic subculture is a stretch. New technology and social networking alone have contributed to poets and other writers' work being shared across the world. YouTube has emerged to be a major asset to poets, allowing live performances to be streamed to billions of users. Even now, poetry continues to grow and voice topics that are relevant to the culture of our time. Poetry is not in the spotlight as it may have been in earlier times, but it's still a relevant art form that continues to expand in scope and appeal.

Furthermore, Gioia's argument does not account for live performances of poetry. Not everyone has taken a poetry class or enrolled in university—but most everyone is online. The Internet is a perfect launching point to get all creative work out there. An example of this was the performance of Buddy Wakefield's *Hurling Crowbirds at Mockingbars*. Wakefield is a well-known poet who has published several collections of contemporary poetry. One of my favorite works by Wakefield is *Crowbirds*, specifically his performance at New York University in 2009. Although his reading was a campus event, views of his performance online number in the thousands. His poetry attracted people outside of the university setting.

Naturally, the poem's popularity can be attributed both to Wakefield's performance and the quality of his writing. *Crowbirds* touches on themes of core human concepts such as faith, personal loss, and growth. These are not ideas that only poets or students of literature understand, but all human beings: "You acted like I was hurling crowbirds at mockingbars / and abandoned me for not making sense. / Evidently, I don't experience things as rationally as you do" (Wakefield 15-17). Wakefield weaves together a complex description of the perplexed and hurt emotions of the speaker undergoing a separation from a romantic interest. The line "You acted like I was hurling crowbirds at mockingbars" conjures up an image of someone confused, seemingly out of their mind . . . or in the case of the speaker, passionately trying to grasp at a relationship that is fading. The speaker is looking back and finding the words that described how he wasn't making sense. This poem is particularly human and gripping in its message, but the entire effect of the poem is enhanced through the physical performance.

At its core, poetry is about addressing issues/ideas in the world. Part of this is also addressing the perspectives that are exiguously considered. Although the platform may look different, poetry continues to have a steady audience due to the emotional connection the poet shares with the audience.

Write an essay in which you explain how the author builds an argument to persuade the audience that poetry is not dying. In your essay, analyze how the author uses one or more of the features listed in the bullet points above (or features of your own choice) to strengthen the logic and persuasiveness of their argument. Focus on the most relevant features of the passage.

Answer Explanations for Practice Test #3

Reading Test

1. D: It emphasizes Mr. Utterson's anguish in failing to identify Hyde's whereabouts. Context clues indicate that Choice *D* is correct because the passage provides great detail of Mr. Utterson's feelings about locating Hyde. Choice *A* does not fit because there is no mention of Mr. Lanyon's mental state. Choice *B* is incorrect; although the text does make mention of bells, Choice *B* is not the *best* answer overall. Choice *C* is incorrect because the passage clearly states that Mr. Utterson was determined, not unsure.

2. A: In the city. The word *city* appears in the passage several times, thus establishing the location for the reader.

3. B: It scares children. The passage states that the Juggernaut causes the children to scream. Choices *A* and *D* don't apply because the text doesn't mention either of these instances specifically. Choice *C* is incorrect because there is nothing in the text that mentions space travel.

4. B: To constantly visit. The mention of *morning*, *noon*, and *night* make it clear that the word *haunt* refers to frequent appearances at various locations. Choice *A* doesn't work because the text makes no mention of levitating. Choices *C* and *D* are not correct because the text makes mention of Mr. Utterson's anguish and disheartenment because of his failure to find Hyde but does not make mention of Mr. Utterson's feelings negatively affecting anyone else.

5. D: This is an example of alliteration. Choice *D* is the correct answer because of the repetition of the *L*-words. Hyperbole is an exaggeration, so Choice *A* doesn't work. No comparison is being made, so no simile or metaphor is being used, thus eliminating Choices *B* and *C*.

6. D: The speaker intends to continue to look for Hyde. Choices *A* and *B* are not possible answers because the text doesn't refer to any name changes or an identity crisis, despite Mr. Utterson's extreme obsession with finding Hyde. The text also makes no mention of a mistaken identity when referring to Hyde, so Choice *C* is also incorrect.

7. A: The author's attitude toward the main subject can be described as *intrigue*. Although this is fiction and we are seeing the passage through the eyes of a character, the author still is in control of word choice and tone. *Intrigue* means to arouse curiosity, so we are confronted with words and phrases such as "besieged by questions," "digging," "imagination," and "haunt." Choice *B*, elation, means joy. Choice *C*, animosity, means strong dislike. Choice *D*, rigidity, means stiff or unyielding.

8. B: Mr. Utterson is struggling with the mystery surrounding a dark figure and the terrible crimes he commits. As Mr. Utterson tosses and turns in bed in the long paragraph, we see him wanting to discern the figure's face as he imagines him committing the crimes, but Mr. Utterson has no idea who the figure is.

9. C: Choice *C* is the best answer because of the chronological aspect of the passage. By the transition "From that time forward," we know that Mr. Utterson haunted storefronts *after* his night visions of the mysterious figure, and we also see that Mr. Utterson's dialogue at the very end is a promise to find "Mr. Hyde," whoever he may be.

10. C: The passage is in third person limited, which means we see the thoughts of one character only by the use of the pronouns "he" or "she." First person is characterized by the use of "I." Second person is characterized by the use of "you." Third person omniscient is when we see the thoughts of all the characters in the story, and the author uses the pronouns "he" or "she."

11. A: The purpose is to inform the reader about what assault is and how it is committed. Choice *B* is incorrect because the passage does not state that assault is a lesser form of lethal force, only that an assault can use lethal force, or alternatively, lethal force can be utilized to counter a dangerous assault. Choice *C* is incorrect because the passage is informative and does not have a set agenda. Finally, Choice *D* is incorrect because although the author uses an example in order to explain assault, it is not indicated that this is the author's personal account.

12. C: If the man being attacked in an alley by another man with a knife used self-defense by lethal force, it would not be considered illegal. The presence of a deadly weapon indicates mal-intent and because the individual is isolated in an alley, lethal force in self-defense may be the only way to preserve his life. Choices *A* and *B* can be ruled out because in these situations, no one is in danger of immediate death or bodily harm by someone else. Choice *D* is an assault and does exhibit intent to harm, but this situation isn't severe enough to merit lethal force; there is no intent to kill.

13. B: As discussed in the second passage, there are several forms of assault, like assault with a deadly weapon, verbal assault, or threatening posture or language. Choice *A* is incorrect because the author does mention what the charges are on assaults; therefore, we cannot assume that they are more or less than unnecessary use of force charges. Choice *C* is incorrect because anyone is capable of assault; the author does not state that one group of people cannot commit assault. Choice *D* is incorrect because assault is never justified. Self-defense resulting in lethal force can be justified.

14. D: The use of lethal force is not evaluated on the intent of the user, but rather on the severity of the primary attack that warranted self-defense. This statement most undermines the last part of the passage because it directly contradicts how the law evaluates the use of lethal force. Choices *A* and *B* are stated in the paragraph, so they do not undermine the explanation from the author. Choice *C* does not necessarily undermine the passage, but it does not support the passage either. It is more of an opinion that does not offer strength or weakness to the explanation.

15. C: An assault with deadly intent can lead to an individual using lethal force to preserve their well-being. Choice *C* is correct because it clearly establishes what both assault and lethal force are and gives the specific way in which the two concepts meet. Choice *A* is incorrect because lethal force doesn't necessarily result in assault. This is also why Choice *B* is incorrect. Not all assaults would necessarily be life-threatening to the point where lethal force is needed for self-defense. Choice *D* is compelling but ultimately too vague; the statement touches on aspects of the two ideas but fails to present the concrete way in which the two are connected to each other.

16. A: Both passages open by defining a legal concept and then continue to describe situations in order to further explain the concept. Choice *D* is incorrect because while the passages utilize examples to help explain the concepts discussed, the author doesn't indicate that they are specific court cases. It's also clear that the passages don't open with examples, but instead, they begin by defining the terms addressed in each passage. This eliminates Choice *B*, and ultimately reveals Choice *A* to be the correct answer. Choice *A* accurately outlines the way both passages are structured. Because the passages follow a nearly identical structure, the Choice *C* can easily be ruled out.

17. C: Intent is very important for determining both lethal force and assault; intent is examined in both parties and helps determine the severity of the issue. Choices *A* and *B* are incorrect because it is clear in both passages that intent is a prevailing theme in both lethal force and assault. Choice *D* is compelling, but if a person uses lethal force to defend himself or herself, the intent of the defender is also examined in order to help determine if there was excessive force used. Choice *C* is correct because it states that intent is important for determining both lethal force and assault, and that intent is used to gauge the severity of the issues. Remember, just as lethal force can escalate to excessive use of force, there are different kinds of assault. Intent dictates several different forms of assault.

18. B: The example is used to demonstrate a single example of two different types of assault, then adding in a third type of assault to the example's conclusion. The example mainly provides an instance of "threatening body language" and "provocative language" with the homeowner gesturing threats to his neighbor. It ends the example by adding a third type of assault: physical strikes. This example is used to show the variant nature of assaults. Choice *A* is incorrect because it doesn't mention the "physical strike" assault at the end and is not specific enough. Choice *C* is incorrect because the example does not say anything about the definition of lethal force or how it might be altered. Choice *D* is incorrect, as the example mentions nothing about cause and effect.

19. C: The word *apprehension* most nearly means fear. The passage indicates that "assault is the unlawful attempt of one person to apply fear/anxiety on another individual by an imminent threat." The creation of fear in another individual seems to be a property of assault.

20. D: Choice *D* is the best answer, "The concept of lethal force as a last resort and the point at which it can cross a line from defense to manslaughter." The last paragraph of the first passage states what the term "last resort" means and how it's distinguished in the eyes of the law.

21. A: The word *patronage* most nearly means *auspices*, which means *protection* or *support*. Choice *B*, *aberration*, means *deformity* and does not make sense within the context of the sentence. Choice *C*, *acerbic*, means *bitter* and also does not make sense in the sentence. Choice *D*, *adulation*, is a positive word meaning *praise*, and thus does not fit with the word *condescending* in the sentence.

22. D: *Working man* is most closely aligned with Choice *D*, *bourgeois*. In the context of the speech, the word *bourgeois* means *working* or *middle class*. Choice *A*, *Plebeian*, does suggest *common people*; however, this is a term that is specific to ancient Rome. Choice *B*, *viscount*, is a European title used to describe a specific degree of nobility. Choice *C*, *entrepreneur*, is a person who operates their own business.

23. C: In the context of the speech, the term *working man* most closely correlates with Choice *C*, "A working man is someone who works for wages among the middle class." Choice *A* is not mentioned in the passage and is off-topic. Choice *B* may be true in some cases, but it does not reflect the sentiment described for the term *working man* in the passage. Choice *D* may also be arguably true. However, it is not given as a definition but as *acts* of the working man, and the topics of *field, factory,* and *screen* are not mentioned in the passage.

24. D: *Enterprise* most closely means *cause*. Choices *A, B,* and *C* are all related to the term *enterprise*. However, Dickens speaks of a *cause* here, not a company, courage, or a game. "He will stand by such an enterprise" is a call to stand by a cause to enable the working man to have a certain autonomy over his own economic standing. The very first paragraph ends with the statement that the working man "shall . . . have a share in the management of an institution which is designed for his benefit."

25. B: The speaker's salutation is one from an entertainer to his audience and uses the friendly language to connect to his audience before a serious speech. Recall in the first paragraph that the speaker is there to "accompany [the audience] . . . through one of my little Christmas books," making him an author there to entertain the crowd with his own writing. The speech preceding the reading is the passage itself, and, as the tone indicates, a serious speech addressing the "working man." Although the passage speaks of employers and employees, the speaker himself is not an employer of the audience, so Choice A is incorrect. Choice C is also incorrect, as the salutation is not used ironically, but sincerely, as the speech addresses the well-being of the crowd. Choice D is incorrect because the speech is not given by a politician, but by a writer.

26. B: Choice A is incorrect because that is the speaker's *first* desire, not his second. Choices C and D are tricky because the language of both of these is mentioned after the word *second*. However, the speaker doesn't get to the second wish until the next sentence. Choices C and D are merely prepositions preparing for the statement of the main clause, Choice B, for the working man to have a say in his institution, which is designed for his benefit.

27. C: The speaker's tone can best be described as *confident and informed.* The speaker addresses the audience as "My good friends," and says, "I have no fear of being misunderstood," which implies confidence. Additionally, the speaker's knowledge of the proposal and topic can be seen in the text as well, especially in the second paragraph.

28. D: To provide credibility to the working man and share confidence in their ability to take on responsibilities if they are compensated appropriately. The speaker provides credibility by saying "he will stand by such an enterprise with the utmost of his patience," and displays their responsibilities by saying "he will feel his responsibility like an honest man."

29. A: The speaker says to "Erect in Birmingham a great Education Institution, properly educational." Choice B is close, but the speaker uses the name "Temple of Concord" in the passage as a metaphor, so this is incorrect. The other two choices aren't mentioned in the passage.

30. B: The word *antagonism* most nearly means opposition. Choice A, *conformity*, is the opposite of antagonism. Choice C, *affluence*, means abundance. Choice D, *scarcity*, means being deficient in something.

31. D: The use of "I" could serve to have a "hedging" effect, allow the reader to connect with the author in a more personal way, and cause the reader to empathize more with the egrets. However, it doesn't distance the reader from the text, making Choice D the answer to this question.

32. C: The quote provides an example of a warden protecting one of the colonies. Choice A is incorrect because the speaker of the quote is a warden, not a hunter. Choice B is incorrect because the quote does not lighten the mood but shows the danger of the situation between the wardens and the hunters. Choice D is incorrect because there is no humor found in the quote.

33. D: A *rookery* is a colony of breeding birds. Although *rookery* could mean Choice A, houses in a slum area, it does not make sense in this context. Choices B and C are both incorrect, as this is not a place for hunters to trade tools or for wardens to trade stories.

34. B: An important bird colony. The previous sentence is describing "twenty colonies" of birds, so what follows should be a bird colony. Choice A may be true, but we have no evidence of this in the text.

Choice *C* does touch on the tension between the hunters and wardens, but there is no official "Bird Island Battle" mentioned in the text. Choice *D* does not exist in the text.

35. D: To demonstrate the success of the protective work of the Audubon Association. The text mentions several different times how and why the association has been successful and gives examples to back this fact. Choice *A* is incorrect because although the article, in some instances, calls certain people to act, it is not the purpose of the entire passage. There is no way to tell if Choices *B* and *C* are correct, as they are not mentioned in the text.

36. C: To have a better opportunity to hunt the birds. Choice *A* might be true in a general sense, but it is not relevant to the context of the text. Choice *B* is incorrect because the hunters are not studying lines of flight to help wardens, but to hunt birds. Choice *D* is incorrect because nothing in the text mentions that hunters are trying to build homes underneath lines of flight of birds for good luck.

37. C: The passage states in the third paragraph that a year before, "a number of egrets were seen in eastern Massachusetts." Florida is mentioned in the passage as a place where bird colonies reside. The other two locations are not mentioned in the passage.

38. A: The word *commissioned* most nearly means *appointed*. Choice *B*, *compelled*, means forced. Choice *C*, *beguiled*, means entertained. Choice *D*, fortified, means defended.

39. D: Several hidden colonies were found which prompted Dutcher and Pearson to protect them. This information is presented in the first paragraph starting with the sentence "Two years ago." The other answer choices are current to the passage.

40. C: *In transit* means on the way to the destination. Choice *A* is the definition of *travelling*. Choice *B* is the definition of *delegate*. Choice *D* is the definition of *provoke*.

41. A: It introduces certain insects that transition from water to air. Choice *B* is incorrect because although the passage talks about gills, it is not the central idea of the passage. Choices *C* and *D* are incorrect because the passage does not "define" or "invite," but only serves as an introduction to stoneflies, dragonflies, and mayflies and their transition from water to air.

42. C: The act of shedding part or all of the outer shell. Choices *A*, *B*, and *D* are incorrect. The word in the passage is mentioned here: "But it is advisable in connection with our present subject to dwell especially on some insects that remain continually under water till they are ready to undergo their final moult and attain the winged state, which they pass entirely in the air."

43. B: The first paragraph serves as a contrast to the second. Notice how the first paragraph goes into detail describing how insects are able to breathe air. The second paragraph acts as a contrast to the first by stating "[i]t is of great interest to find that, nevertheless, a number of insects spend much of their time under water." Watch for transition words such as "nevertheless" to help find what type of passage you're dealing with.

44: C: The stage of preparation in between molting is acted out in the water, while the last stage is in the air. Choices *A*, *B*, and *D* are all incorrect. *Instars* is the phase between two periods of molting, and the text explains when these transitions occur.

45. C: The author's tone is informative and exhibits interest in the subject of the study. Overall, the author presents us with information on the subject. One moment where personal interest is depicted is when the author states, "It is of great interest to find that, nevertheless, a number of insects spend much of their time under water."

46. C: Their larva can breathe the air dissolved in water through gills of some kind. This is stated in the last paragraph. Choice *A* is incorrect because the text mentions this in a general way at the beginning of the passage concerning "insects as a whole." Choice *B* is incorrect because this is stated of beetles and water-bugs, and not the insects in question. Choice *D* is incorrect because this is the opposite of what the text says of instars.

47. B: According to the passage, boatmen and scorpions have some way of protecting their spiracles when submerged. We see this in the second paragraph, which says "(boatmen and scorpions) which have some way of protecting their spiracles when submerged."

48. A: The best answer Choice is *A*: the author believes that the stonefly, dragonfly, and mayfly larvae are better prepared to live beneath the water because they have gills that allow them to do so. We see this when the author says "But the larva of a stone-fly, a dragon-fly, or a may-fly is adapted more completely than these for aquatic life; it can, by means of gills of some kind, breathe the air dissolved in water."

49. C: Because a vast majority of insects have wings and also have the ability to breathe underwater. The entire first paragraph talks of how insects have wings, and how insects also have "a system of branching air-tubes" that carries oxygen to the insect's tissues.

50. B: The word *preeminently* most nearly means *above all* or *in particular*. The author is saying that above all, insects are creatures of both land and water.

Writing and Language Test

1. D: This passage is informative because it is nonfiction and factual. The passage's intent is not to state an opinion, discuss an individual's life, or tell a story. Thus, the passage is not argumentative (Choice *A*), biographical (Choice *C*), or narrative (Choice *B*).

2. D: To begin, *of* is not required here. *Apprenticeship* is also more appropriate in this context than *apprentice opportunities*, *apprentice* describes an individual in an apprenticeship, not an apprenticeship itself. Both of these changes are needed, making Choice *D* the correct answer.

3. D: To begin, the selected sentence is a run-on, and displays confusing information. Thus, the sentence does need revision, making (Choice *A*) wrong. The main objective of the selected section of the passage is to communicate that many positions (*positions* is a more suitable term than *offices*, as well) require a PE license, which is gained by scoring well on the FE and PE assessments. This must be the primary focus of the revision. It is necessary to break the sentence into two, to avoid a run-on. Choice *B* fixes the run-on aspect, but the sentence is indirect and awkward in construction. It takes too long to establish the importance of the PE license. Choice *C* is wrong for the same reason and it is a run on. Choice *D* is correct because it breaks the section into coherent sentences and emphasizes the main point the author is trying to communicate: the PE license is required for some higher positions, it's obtained by scoring well on the two standardized assessments, and college and experience can be used to prepare for the assessments in order to gain the certification.

4. C: Any time a writer wants to validate a claim, he or she ought to provide factual information that proves or supports that claim: "beginning his or her preparation for the profession in high school" supports the claim that aircraft engineers undergo years of education. For this reason, Choice *C* is the correct response. However, completing such courses in high school does not guarantee that aircraft engineers will earn generous salaries (Choice *A*), become employed in executive positions (Choice *B*), or stay employed (Choice *D*).

5. B: Choice *B* is correct because *skill* is defined as having certain aptitude for a given task. (Choice *C*) is incorrect because *work* does not directly denote "critical thinking, business skills, problem solving, and creativity." (*A*) is incorrect because the word *care* doesn't fit into the context of the passage, and (*D*), *composition*, is incorrect because nothing in this statement points to the way in which something is structured.

6. C: *Allows* is inappropriate because it does not stress what those in the position of aircraft engineers actually need to be able to do. *Requires* is the only alternative that fits because it actually describes necessary skills of the job.

7. B: The third paragraph discusses reports made by the United States Bureau of Labor Statistics (BLS) in regards to the median, upper 10 percent, and lower 10 percent annual salaries of aircraft engineers in 2015. Therefore, this paragraph is used to support the claim that aircraft engineers are compensated with generous salaries (Choice *B*). The paragraph has nothing to do with an aircraft engineer's skill set (Choice *A*), education (Choice *C*), or incentive program (Choice *D*).

8. A: The correct response is Choice *A* because this statement's intent is to give examples as to how aircraft engineers apply mathematical equations and scientific processes towards aeronautical and aerospace issues and/or inventions. The answer is not *therefore* (Choice *B*) or *furthermore* (Choice *D*) because no causality is being made between ideas. Two items are neither being compared nor contrasted, so *however* (Choice *C*) is also not the correct answer.

9. A: No change is required. The comma is properly placed after the introductory phrase "In May of 2015." Choice *B* is missing the word "in." Choice *C* does not separate the introductory phrase from the rest of the sentence. Choice *D* places an extra, and unnecessary, comma prior to 2015.

10. A: The word *conversely* best demonstrates the opposite sentiments in this passage. Choice *B* is incorrect because it denotes agreement with the previous statement. Choice *C* is incorrect because the sentiment is not restated but opposed. Choice *D* is incorrect because the previous statement is not a cause for the sentence in question.

11. A: Choice *A* is the correct answer because the projections are taking place in the present, even though they are making reference to a future date.

12. B: The passage contains clearly labeled subheadings. These subheadings inform the reader what will be addressed in upcoming paragraphs. Choice *A* is incorrect because the anti-terrorism laws of other countries were never addressed in the passage. The text is written in an informative manner; overly descriptive language is not utilized. Therefore, Choice *C* is incorrect. Choice *D* is incorrect because as mentioned, the structure of the text does help in the manner described in Choice *B*.

13. A: No change is needed. Choices *B* and *C* have incorrect comma placements. Choice *D* utilizes an incorrect verb tense (*responding*).

14. D: The third paragraph states "The majority of Muslims practice Islam peacefully". Therefore, the author explicitly states that most Muslims are peaceful peoples (Choice *D*). Choices *A*, *B*, and *C*, are not included in the passage and are incorrect.

15. B: The term *violate* implies a lack of respect or compliance. *Defile* means to degrade or show no respect. Therefore, Choice *B* is the correct answer. Choice *A* is incorrect because *respect* is the opposite of violate. To "deny" is to refuse, so Choice *C* is not the answer because the weight of the word *deny* is not as heavy as the word *violate*. To "obey" is to follow orders, so Choice *D* is also incorrect.

16. B: An allusion is a direct or indirect literary reference or figure of speech towards a person, place, or event. By referencing the diversion of the airplanes to alternate locations, the author uses an allusion (Choice *B*) to highlight the impact of United Flight 93. Although a graph depicting the decline in the number of aircraft passengers is provided, an image is not. Therefore, Choice *A* is not the answer. The passage does not tell the story from a single passenger's point of view. Thus, Choices *C* and *D* are not the answers.

17. C: All of the choices except *C* go with the flow of the original underlined portion of the sentence and communicate the same idea. Choice *C*, however, does not take into account the rest of the sentence and therefore, becomes awkward and incorrect.

18. B: Although *diverging* means to separate from the main route and go in a different direction, it is used awkwardly and unconventionally in this sentence. Therefore, Choice *A* is not the answer. Choice *B* is the correct answer because it implies that the passengers distracted the terrorists, which caused a change in the plane's direction. Choice *C*, *converging*, is incorrect because it implies that the plane met another in a central location. Although the passengers may have distracted the terrorists, they did not distract the plane. Therefore, Choice *D* is incorrect.

19. D: The graph shows the number of people (in millions) boarding United States' flights between 1996-2005. The first few months following the attacks, the passengers boarding U.S. flights dropped to around 50 million when before the attacks there were around 70 million passengers boarding flights. Therefore, the correct answer is Choice *D*. The graph does not show where the flights were redirected (Choice *B*), the number of passengers that other countries received as a result of the redirected air travel (Choice *C*), or the resulting flight schedule implications (Choice *A*).

20. B: The last paragraph explains that museums and monuments have been erected to honor those who died as a result of the attacks and those who risked their lives to save the injured. Thus, the paragraph serves to explain the lasting impact on America and honor those impacted by the event (Choice *B*). The design of the museums and monuments are not described, so Choice *A* is incorrect. Choice *C* is incorrect because America's War on Terror was not discussed in the last paragraph. Choice *D* is incorrect, because although the previous location of the towers was converted into a park, this was not mentioned in the passage.

21. C: *Intended* means "planned" or "meant to". *Intended* is a far better choice than *desired*, because it would communicate goals and strategy more than simply saying that Bush desired to do something. *Desired* communicates wishing or direct motive. Choices *B* and *D* have irrelevant meanings and wouldn't serve the sentence at all.

22. A: While Choice *B* isn't necessarily wrong, it lacks the direct nature that the original sentence has. Also, by breaking up the sentences like this, the reader becomes confused because the connection between the Taliban's defeat and ongoing war is now separated by a second sentence that is not necessary. Choice *C* corrects this problem but the fluidity of the sentence is marred because of the awkward construction of the first sentence. Choice *D* begins well, but lacks the use of *was* before *overthrown*, which discombobulates the sentence. While *yet* provides an adequate transition for the next sentence, the term *however* is more appropriate. Thus, the original structure of the two sentences is correct, making Choice *A* the best answer.

23. A: The comma after *result* is necessary for the sentence structure, making it an imperative component. The original sentence is correct, making Choice *A* correct. For the reason just listed, Choice *B* is incorrect because it lacks the crucial comma that introduces a new idea. Choice *C* is incorrect because a colon is unnecessary, and Choice *D* is wrong because the addition of the preposition *of* is both unnecessary and incorrect when applied to the rest of the sentence.

24. C: To be "gifted" is to be talented. *Academically* refers to education. Therefore, Fred Hampton was intellectually talented, or intelligent (Choice *C*). Choice *B* is incorrect because it refers to a level of energy or activity. Choice *A* is incorrect because *vacuous* means the opposite of being gifted academically. Choice *D* is incorrect because it refers to one's physical build and/or abilities.

25. C: The goal for this question is to select a sentence that not only affirms, or backs up, the selected statement, but could also appear after it and flow with the rest of the piece. Choice *A* is irrelevant to the sentence; just because new members earned scholarships this doesn't necessarily mean that this was testament of Hampton's leadership or that this actually benefitted the NAACP. Choice *B* is very compelling. If Hampton got an award for the increase in numbers, this could bolster the idea that he was the direct cause of the rise in numbers and that he was of great value to the organization. However, it does not say directly that he was the cause of the increase and that this was extremely beneficial to the NAACP. Choice *C* is a much better choice than Choice *B* because it mentions that the increase in members is unprecedented. Because there has never been this large an increase before, it can be concluded that this increase was most likely due to Hampton's contributions. Thus, Choice *C* is correct. Choice *D* does nothing for the underlined section.

26. B: Choice *B* moves the word *eventually* to the beginning of the sentences. By using the term as an introductory word, continuity from one sentence to another is created. Meanwhile, the syntax is not lost. Choice *A* is incorrect because the sentence requires a proper transition. Choice *C* is incorrect because the sentence does not contain surprising or contrasting information, as is indicated by the introductory word *nevertheless*. Choice *D* is incorrect because the term *then* implies that Hampton's relocation to the BPP's headquarters in Chicago occurred shortly or immediately after leading the NAACP.

27. D: An individual with a charismatic personality is charming and appealing to others. Therefore, Choice *D* is the correct answer. Choice *A* is incorrect because someone with an egotistical personality is conceited or self-serving. Choice *B* is incorrect because *obnoxious* is the opposite of charismatic. Choice *C* is incorrect because someone with a chauvinistic personality is aggressive or prejudiced towards one's purpose, desire, or sex.

28. A: The sentence is correct as-is, therefore Choice *A* is correct. The list of events accomplished by Hampton is short enough that each item in the list can be separated by a comma. Choice *B* is incorrect. Although a colon can be used to introduce a list of items, it is not a conventional choice for separating items within a series. Semicolons are used to separate at least three items in a series that have an internal comma. Semicolons can also be used to separate clauses in a sentence that contain internal commas intended for clarification purposes. Neither of the two latter uses of semicolons is required in the example sentence. Therefore, Choice *C* is incorrect. Choice *D* is incorrect because a dash is not a conventional choice for punctuating items in a series.

29. D: Claims can be supported with evidence or supporting details found within the text. Choice *D* is correct because Choices *A*, *B*, and *C* are all either directly stated or alluded to within the passage.

30. A: The term *neutralize* means to counteract, or render ineffective, which is exactly what the FBI is wanting to do. Accommodate means to be helpful or lend aid, which is the opposite of *neutralize*. Therefore, Choice *B* is wrong. *Assuage* means to ease, while *praise* means to express warm feeling, so they are in no way close to the needed context. Therefore, *neutralize* is the best option, making Choice *A* the best answer.

31. A: The original sentence suggests that the floor plans were provided to the FBI by O'Neal which facilitated the identification of the exact location of Hampton's bed. Choice *B* is incorrect because this states that the FBI provided O'Neal with the floor plans instead of the other way around. Choice *C* is incorrect because the sentence's word order conveys the meaning that O'Neal provided the FBI with Hampton's bed as well as the floor plans. Choice *D* is incorrect because it implies that it was the location of the bed that provided the FBI with the headquarters' floor plans.

32. B: *Commemorates* means to honor, celebrate, or memorialize a person or event. Therefore, Choice *B* is correct. Choice *A* is incorrect because *disregards* is the opposite of *commemorates*. Choice *C* is incorrect because *to communicate* means to converse or to speak. Choice *D* is incorrect because *to deny* means to reject, negate, refuse, or rebuff.

33. D: From the context of the passage, it is clear that the author does not think well of the FBI and their investigation of Hampton and the Black Panthers. Choices *B* and *C* can be easily eliminated. "Well intended" is positive, which is not a characteristic that he would probably attribute to the FBI in the passage. Nor would he think they were "confused" but deliberate in their methods. Choice *A, corrupt,* is very compelling; he'd likely agree with this, but Choice *D, prejudiced* is better. The FBI may not have been corrupt, but there certainly seemed to be a particular dislike or distrust for the Black Panthers. Thus, Choice D, *prejudiced*, is correct.

34. D: All of the cities included in the graphs are along the East Coast of the United States. All of the bars on the graphs show an increase in sea level or the number of days with flood events since 1970. Therefore, the author chose to include the graphs to support the claim that sea levels have risen along the East Coast since 1970, Choice *D*.

Choices *A* and *B* are incorrect because the bars above 1970 on Boston's and Atlantic City's graphs are longer than the graphs' bars above 1980. Therefore, during 1970–1980, Boston and Atlantic City both experienced an increase in the number of days with flood events. It's important to note that while there was a decrease from one decade to another, it does not negate the overall trend of increase in flooding events. Choice *C* is incorrect because the bars increase in height on all of the cities' graphs, showing an increase in the number of days with floods along the entire East coast.

35. B: Although *affect* and *effect* sound the same, they have different meanings. *Affect* is used as a verb. It is defined as the influence of a person, place, or event on another. *Effect* is used as a noun. It is defined as the result of an event. Therefore, the latter ought to be used in the heading. For this reason, Choices *C* and *D* are incorrect. Because the effect is a result of the flood, a possessive apostrophe is needed for the singular noun *flood*. For this reason, Choice *A* is incorrect and Choice *B* is correct.

36. D: Again, the objective for questions like this is to determine if a revision is possible within the choices and if it can adhere to the specific criteria of the question; in this case, we want the sentence to maintain the original meaning while being more concise, or shorter. Choice *B* can be eliminated. The meaning of the original sentence is split into two distinct sentences. The second of the two sentences is also incorrectly constructed. Choice *C* is very intriguing, but there is a jumble of verbs present in "Flooding occurs slowly or rapidly submerging" that make the sentence awkward and difficult to understand without the use of a comma after *rapidly*, making it a poor construction. Choice *C* is wrong. Choice *D* is certainly more concise and it is correctly phrased; it communicates the meaning message that flooding can overtake great lengths of land either slowly or very fast. The use of "Vast areas of land" infers that smaller regions or small areas can flood just as well. Thus, Choice *D* is a good revision that maintains the meaning of the original sentence while being concise and more direct. This rules out Choice *A* in the process.

37. B: In this sentence, the word *ocean* does not require an *s* after it to make it plural because "ocean levels" is plural. Therefore, Choices *A* and *C* are incorrect. Because the passage is referring to multiple— if not all ocean levels—*ocean* does not require an apostrophe (*'s*) because that would indicate that only one ocean is the focus, which is not the case. Choice *D* does not fit well into the sentence and, once again, we see that *ocean* has an *s* after it. This leaves Choice *B*, which correctly completes the sentence and maintains the intended meaning.

38. C: Choice *C* is the best answer because it most closely maintains the sentence pattern of the first sentence of the paragraph, which begins with a noun and passive verb phrase. Choice *B* is incorrect because it does not maintain the sentence pattern of the first sentence of the paragraph. Instead, Choice *B* shifts the placement of the modifying prepositional phrase to the beginning of the sentence. Choice *D* is incorrect because it does not maintain the sentence pattern established by the first sentence of the paragraph. Instead, Choice *D* is an attempt to combine two independent clauses.

39. A: Choice *C* can be eliminated because creating a new sentence with *not* is grammatically incorrect and it throws off the rest of the sentence. Choice *B* is wrong because a comma is definitely needed after *devastation* in the sentence. Choice *D* is also incorrect because *while* is a poor substitute for *although*. *Although* in this context is meant to show contradiction with the idea that floods are associated with devastation. Therefore, none of these choices would be suitable revisions because the original was correct; Choice *A* is the correct answer.

40. B: Choice *B* is the correct answer because the final paragraph summarizes key points from each subsection of the text. Therefore, the final paragraph serves as the conclusion. A concluding paragraph is often found at the end of a text. It serves to remind the reader of the main points of a text. Choice *A* is incorrect because the last paragraph does not just mention adverse effects of floods. For example, the paragraph states, "By understanding flood cycles, civilizations can learn to take advantage of flood seasons." Choice *C* is incorrect; although the subheading mentions the drying of floods, the phenomena is not mentioned in the paragraph. Finally, Choice *D* is incorrect because no new information is presented in the last paragraph of the passage.

41. A: Idea and claims are best expressed and supported within a text through examples, evidence, and descriptions. Choice *A* is correct because it provides examples of rivers that support the tenth paragraph's claim that "not all flooding results in adverse circumstances." Choice *B* is incorrect because the sentence does not explain how floods are beneficial. Therefore, Choices *C* and *D* are incorrect.

42. D: In the sentence, *caused* is an incorrect tense, making Choice *A* wrong. Choice *B* is incorrect because *cause* is used as a noun or imperative verb form, we need *cause* in verb form. Choices *C* and *D* are very compelling. Choice *C*, *causing*, is a verb in the present continuous tense, which appears to agree with the verb flooding, but it is incorrectly used. This leaves Choice *D*, *causes*, which does fit because it is in the indefinite present tense. Fitting each choice into the sentence and reading it in your mind will also reveal that Choice *D*, *causes*, correctly completes the sentence. Apply this method to all the questions when possible.

43. A: To *project* means to anticipate or forecast. This goes very well with the sentence because it describes how new technology is trying to estimate flood activity in order to prevent damage and save lives. *Project* in this case needs to be assisted by *to* in order to function in the sentence. Therefore, Choice *A* is correct. Choices *B* and *D* are the incorrect tenses. Choice *C* is also wrong because it lacks *to*.

44. C: *Picturesque* is an adjective used for an attractive, scenic, or otherwise striking image. Thus, Choice *C* is correct. Choice *A* is incorrect because although *colorful* can be included in a picturesque view, it does not encompass the full meaning of the word. Choice *B* is incorrect because *drab* is the opposite of *picturesque*. Choice *D* is incorrect because *candid* is defined as being frank, open, truthful, or honest.

Math Test

1. B: First, subtract 4 from each side. This yields $6t = 12$. Now, divide both sides by 6 to obtain $t = 2$.

2. B: To be directly proportional means that $y = mx$. If x is changed from 5 to 20, the value of x is multiplied by 4. Applying the same rule to the y-value, also multiply the value of y by 4. Therefore, $y = 12$.

3. B: From the slope-intercept form, $y = mx + b$, it is known that b is the y-intercept, which is 1. Compute the slope as $\frac{2-1}{1-0} = 1$, so the equation should be $y = x + 1$.

4. A: Each bag contributes $4x + 1$ treats. The total treats will be in the form $4nx + n$ where n is the total number of bags. The total is in the form $60x + 15$, from which it is known that $n = 15$.

5. D: Let a be the number of apples and b the number of bananas. Then, the total cost is $2a + 3b = 22$, while it also known that $a + b = 10$. Using the knowledge of systems of equations, cancel the b variables by multiplying the second equation by -3. This makes the equation $-3a - 3b = -30$. Adding this to the first equation, the b values cancel to get $-a = -8$, which simplifies to $a = 8$.

6. A: Finding the roots means finding the values of x when y is zero. The quadratic formula could be used, but in this case it is possible to factor by hand, since the numbers -1 and 2 add to 1 and multiply to -2. So, factor:

$$x^2 + x - 2 = (x - 1)(x + 2) = 0$$

then set each factor equal to zero. Solving for each value gives the values $x = 1$ and $x = -2$.

7. C: To find the y-intercept, substitute zero for x, which gives us:

$$y = 0^{\frac{5}{3}} + (0 - 3)(0 + 1)$$

$$0 + (-3)(1)$$

$$-3$$

8. A: This has the form $t^2 - y^2$, with $t = x^2$ and $y = 4$. It's also known that $t^2 - y^2 = (t + y)(t - y)$, and substituting the values for t and y into the right-hand side gives:

$$(x^2 - 4)(x^2 + 4)$$

9. A: Simplify this to:

$$(4x^2y^4)^{\frac{3}{2}} = 4^{\frac{3}{2}}(x^2)^{\frac{3}{2}}(y^4)^{\frac{3}{2}}$$

Now, simplify the numeric term in the expression:

$$4^{\frac{3}{2}} = (\sqrt{4})^3 = 2^3 = 8$$

For the other, recall that the exponents must be multiplied, so this yields:

$$8x^{2 \times \frac{3}{2}}y^{4 \times \frac{3}{2}} = 8x^3y^6$$

10. B: Start by squaring both sides to get $1 + x = 16$. Then subtract 1 from both sides to get $x = 15$.

11. C: Multiply both sides by x to get $x + 2 = 2x$, which simplifies to $-x = -2$, or $x = 2$.

12. B: The independent variable's coordinate at the vertex of a parabola (which is the highest point, when the coefficient of the squared independent variable is negative) is given by $x = -\frac{b}{2a}$.

Substitute and solve for x to get:

$$x = -\frac{4}{2(-16)} = \frac{1}{8}$$

Using this value of x, the maximum height of the ball (y), can be calculated. Substituting x into the equation yields:

$$h(t) = -16\left(\frac{1}{8}\right)^2 + 4\left(\frac{1}{8}\right) + 6 = 6.25$$

13. D: Denote the width as w and the length as l. Then, $l = 3w + 5$. The perimeter is $2w + 2l = 90$. Substituting the first expression for l into the second equation yields:

$$2(3w + 5) + 2w = 90$$

$$6w + 10 + 2w = 90$$

$$8w = 80$$

$$w = 10$$

Putting this into the first equation, it yields:

$$l = 3(10) + 5 = 35$$

14. A: Lining up the given scores provides the following list: 60, 75, 80, 85, and one unknown. Because the median needs to be 80, it means 80 must be the middle data point out of these five. Therefore, the unknown data point must be the fourth or fifth data point, meaning it must be greater than or equal to 80. The only answer that fails to meet this condition is 60.

15. B: If 60% of 50 workers are women, then there are 30 women working in the office. If half of them are wearing skirts, then that means 15 women wear skirts. Since none of the men wear skirts, this means there are 15 people wearing skirts.

16. A: Let the unknown score be x. The average will be:

$$\frac{5 \times 50 + 4 \times 70 + x}{10} = \frac{530 + x}{10} = 55$$

Multiply both sides by 10 to get $530 + x = 550$, or $x = 20$.

17. D: For manufacturing costs, there is a linear relationship between the cost to the company and the number produced, with a y-intercept given by the base cost of acquiring the means of production, and a slope given by the cost to produce one unit. In this case, that base cost is $50,000, while the cost per unit is $40. So:

$$y = 40x + 50,000$$

18. C: A die has an equal chance for each outcome. Since it has six sides, each outcome has a probability of $\frac{1}{6}$. The chance of a 1 or a 2 is therefore:

$$\frac{1}{6} + \frac{1}{6} = \frac{1}{3}$$

19. A: The slope is given by:

$$m = \frac{y_2 - y_1}{x_2 - x_1}$$

$$\frac{0 - 4}{0 - (-3)}$$

$$-\frac{4}{3}$$

20. C: The expected value is equal to the total sum of each product of individual score and probability. There are 36 possible rolls. The probability of rolling a 2 is $\frac{1}{36}$. The probability of rolling a 3 is $\frac{2}{36}$. The probability of rolling a 4 is $\frac{3}{36}$. The probability of rolling a 5 is $\frac{4}{36}$. The probability of rolling a 6 is $\frac{5}{36}$. The probability of rolling a 7 is $\frac{6}{36}$. The probability of rolling an 8 is $\frac{5}{36}$. The probability of rolling a 9 is $\frac{4}{36}$. The probability of rolling a 10 is $\frac{3}{36}$. The probability of rolling an 11 is $\frac{2}{36}$. Finally, the probability of rolling a 12 is $\frac{1}{36}$.

Each possible outcome is multiplied by the probability of it occurring. Like this:

$$2 \times \frac{1}{36} = a$$

$$3 \times \frac{2}{36} = b$$

$$4 \times \frac{3}{36} = c$$

And so forth.

Then all of those results are added together:

$$a + b + c \ldots = expected\ value$$

In this case, it equals 7.

21. A: The graph contains four turning points (where the curve changes from rising to falling or vice versa). This indicates that the degree of the function (highest exponent for the variable) is 5, eliminating Choices *C* and *D*. The y-intercepts of the functions can be determined by substituting 0 for x and finding the value of y. The function for Choice *A* has a y-intercept of 3, and the function for Choice *B* has a y-intercept of -3. Therefore, Choice *B* is eliminated.

22. C: $\frac{1}{3}$ of the shirts sold were patterned. Therefore, $1 - \frac{1}{3} = \frac{2}{3}$ of the shirts sold were solid. Anytime "of" a quantity appears in a word problem, multiplication should be used. Therefore:

$$192 \times \frac{2}{3}$$

$$\frac{192 \times 2}{3}$$

$$\frac{384}{3}$$

128 solid shirts were sold

The entire expression is:

$$192 \times \left(1 - \frac{1}{3}\right)$$

23. A: Mean. An outlier is a data value that is either far above or far below the majority of values in a sample set. The mean is the average of all the values in the set. In a small sample set, a very high or very low number could drastically change the average of the data points. Outliers will have no more of an effect on the median (the middle value when arranged from lowest to highest) than any other value above or below the median. If the same outlier does not repeat, outliers will have no effect on the mode (value that repeats most often).

24. C: Line graph. The scenario involves data consisting of two variables, month and stock value. Box plots display data consisting of values for one variable. Therefore, a box plot is not an appropriate choice. Both line plots and circle graphs are used to display frequencies within categorical data. Neither can be used for the given scenario. Line graphs display two numerical variables on a coordinate grid and show trends among the variables.

25. D: $\frac{1}{12}$. The probability of picking the winner of the race is $\frac{1}{4}$ or:

$$\left(\frac{number\ of\ favorable\ outcomes}{number\ of\ total\ outcomes}\right)$$

Assuming the winner was picked on the first selection, three horses remain from which to choose the runner-up (these are dependent events). Therefore, the probability of picking the runner-up is $\frac{1}{3}$. To determine the probability of multiple events, the probability of each event is multiplied:

$$\frac{1}{4} \times \frac{1}{3} = \frac{1}{12}$$

26. C: Each number in the sequence is adding one more than the difference between the previous two.

For example, $10 - 6 = 4, 4 + 1 = 5$.

Therefore, the next number after 10 is $10 + 5 = 15$.

Going forward, $21 - 15 = 6, 6 + 1 = 7$. The next number is $21 + 7 = 28$. Therefore, the difference between numbers is the set of whole numbers starting at 2: 2, 3, 4, 5, 6, 7....

27. D: The formula for finding the volume of a rectangular prism is $V = l \times w \times h$ where l is the length, w is the width, and h is the height. The volume of the original box is calculated:

$$V = 8 \times 14 \times 4 = 448 \text{ in}^3$$

The volume of the new box is calculated:

$$V = 16 \times 28 \times 8 = 3584 \text{ in}^3$$

The volume of the new box divided by the volume of the old box equals 8.

28. A: The notation i stands for an imaginary number. The value of i is equal to $\sqrt{-1}$. When performing calculations with imaginary numbers, treat i as a variable, and simplify when possible. Multiplying the binomials by the FOIL method produces:

$$15 - 12i + 10i - 8i^2$$

Combining like terms yields:

$$15 - 2i - 8i^2$$

Since $i = \sqrt{-1}, i^2 = (\sqrt{-1})^2 = -1$.

Therefore, substitute -1 for i^2:

$$15 - 2i - 8(-1)$$

Simplifying results in:

$$15 - 2i + 8 = 23 - 2i$$

29. C: The formula to find arc length is $s = \theta r$ where s is the arc length, θ is the radian measure of the central angle, and r is the radius of the circle. Substituting the given information produces: 3π cm $= \theta 12$ cm. Solving for θ yields $\theta = \frac{\pi}{4}$. To convert from radian to degrees, multiply the radian measure by $\frac{180°}{\pi}$:

$$\frac{\pi}{4} \times \frac{180°}{\pi} = 45°$$

30. B: Given the area of the circle, the radius can be found using the formula $A = \pi r^2$. In this case, $49\pi = \pi r^2$, which yields $r = 7$ m. A central angle is equal to the degree measure of the arc it inscribes; therefore, $\angle x = 80°$. The area of a sector can be found using the formula:

$$A = \frac{\theta}{360°} \times \pi r^2$$

In this case:

$$A = \frac{80°}{360°} \times \pi(7)^2 = 10.9\pi \text{ m}$$

31. D: SOHCAHTOA is used to find the missing side length. Because the angle and adjacent side are known, $\tan 60 = \frac{x}{13}$.

Making sure to evaluate tangent with an argument in degrees, this equation gives"

$$x = 13 \tan 60 = 13 \times \sqrt{3} = 22.52$$

32. C: The sample space is made up of $8 + 7 + 6 + 5 = 26$ balls.

The probability of pulling each individual ball is $\frac{1}{26}$. Since there are 7 yellow balls, the probability of pulling a yellow ball is $\frac{7}{26}$.

33. D: The addition rule is necessary to determine the probability because a 6 can be rolled on either roll of the die. The rule used is:

$$P(A \text{ or } B) = P(A) + P(B) - P(A \text{ and } B)$$

The probability of a 6 being individually rolled is $\frac{1}{6}$ and the probability of a 6 being rolled twice is:

$$\frac{1}{6} \times \frac{1}{6} = \frac{1}{36}$$

Therefore, the probability that a 6 is rolled at least once is:

$$\frac{1}{6} + \frac{1}{6} - \frac{1}{36} = \frac{11}{36}$$

34. A: If each man gains 10 pounds, every original data point will increase by 10 pounds. Therefore, the man with the original median will still have the median value, but that value will increase by 10. The smallest value and largest value will also increase by 10 and, therefore, the difference between the two won't change. The range does not change in value and, thus, remains the same.

35. D: When an ordered pair is reflected over an axis, the sign of one of the coordinates must change. When it's reflected over the x-axis, the sign of the y-coordinate must change. The x-value remains the same. Therefore, the new ordered pair is $(-3, 4)$.

36. A: Because the volume of the given sphere is 288π cubic meters, this gives:

$$\frac{4}{3}\pi r^3 = 288\pi$$

This equation is solved for r to obtain a radius of 6 meters. The formula for surface area is $4\pi r^2$ so:

$$SA = 4\pi 6^2 = 144\pi \text{ square meters}$$

37. B: A rectangle is a specific type of parallelogram. It has 4 right angles. A square is a rhombus that has 4 right angles. Therefore, a square is always a rectangle because it has two sets of parallel lines and 4 right angles.

38. B: The sine of 30° is equal to $\frac{1}{2}$. Choice *A* is not the correct answer because the sine of 15° is 0.2588. Choice *C* is not the answer because the sine of 45° is 0.7071. Choice *D* is not the answer because the sine of 90 degrees is 1.

39. C: The cosine of 45° is equal to 0.7071. Choice *A* is not the correct answer because the cosine of 15° is 0.9659. Choice *B* is not the correct answer because the cosine of 30° is 0.8660. Choice *D* is not correct because the cosine of 90° is 0.

40. B: The tangent of 30° is 1 over the square root of 3. Choice *A* is not the correct answer because the tangent of 15° is 0.2679. Choice *C* is not the correct answer because the tangent of 45° is 1. Choice *D* is not the correct answer because the tangent of 90° is undefined.

41. C: Graphing the function $y = \cos(x)$ shows that the curve starts at $(0, 1)$, has an amplitude of 2, and a period of 2π. This same curve can be constructed using the sine graph, by shifting the graph to the left $\frac{\pi}{2}$ units. This equation is in the form $y = \sin\left(x + \frac{\pi}{2}\right)$.

42. A: Every 8 mL of medicine requires 5 mL. The 45 mL first needs to be split into portions of 8 mL. This results in $\frac{45}{8}$ portions. Each portion requires 5 mL. Therefore, $\frac{45}{8} \times 5 = \frac{45 \times 5}{8} = \frac{225}{8}$ mL is necessary.

43. D: The midpoint formula should be used to get the average of both points.

$$M = \left(\frac{x_1 + x_2}{2}, \frac{y_1 + y_2}{2}\right)$$

$$\left(\frac{-1 + 3}{2}, \frac{2 + (-6)}{2}\right) = (1, -2)$$

44. A: First, the sample mean must be calculated.

$$\bar{x} = \frac{1}{4}(1 + 3 + 5 + 7) = 4$$

The sample standard deviation of the data set is:

$$s = \sqrt{\frac{\sum(x - \bar{x})^2}{n - 1}}$$

and $n = 4$ represents the number of data points.

Therefore, the sample standard deviation is:

$$s = \sqrt{\frac{1}{3}[(1 - 4)^2 + (3 - 4)^2 + (5 - 4)^2 + (7 - 4)^2]}$$

$$s = \sqrt{\frac{1}{3}(9 + 1 + 1 + 9)} = 2.58$$

45. B: An equilateral triangle has three sides of equal length, so if the total perimeter is 18 feet, each side must be 6 feet long. A square with sides of 6 feet will have an area of $6^2 = 36$ square feet.

46. A: The formula for the volume of a sphere is $\frac{4}{3}\pi r^3$, and $\frac{4}{3} \times \pi \times 3^3$ is 36π in³. Choice B is not the correct answer because that is only 3^3. Choice C is not the correct answer because that is 3^2 and Choice D is not the correct answer because that is 36×2.

47. A: This answer is correct because $100 - 64$ is 36, and taking the square root of 36 is 6. Choice B is not the correct answer because that is $10 + 8$. Choice C is not the correct answer because that is 8×10. Choice D is also not the correct answer because there is no reason to arrive at that number.

48. A: The formula for the area of the circle is πr^2 and 9 squared is 81. Choice B is not the correct answer because that is 2×9. Choice C is not the correct answer because that is 9×10. Choice D is not the correct answer because that is simply the value of the radius.

49. B: 13,078. The power of 10 by which a digit is multiplied corresponds with the number of zeros following the digit when expressing its value in standard form. Therefore:

$$(1 \times 10^4) + (3 \times 10^3) + (7 \times 10^1) + (8 \times 10^0)$$

$$10,000 + 3,000 + 70 + 8 = 13,078$$

50. A: Using the trigonometric identity $\tan(\theta) = \frac{\sin(\theta)}{\cos(\theta)}$, the expression becomes $\frac{\sin\theta}{\cos\theta}\cos\theta$. The factors that are the same on the top and bottom cancel out, leaving the simplified expression $\sin\theta$.

51. B: $\frac{11}{15}$. Fractions must have like denominators to be added. We are trying to add a fraction with a denominator of 3 to a fraction with a denominator of 5, so we have to convert both fractions to their respective equivalent fractions that have a common denominator. The common denominator is the least common multiple (LCM) of the two original denominators. In this case, the LCM is 15, so both fractions should be changed to equivalent fractions with a denominator of 15.

To determine the numerator of the new fraction, the old numerator is multiplied by the same number by which the old denominator is multiplied to obtain the new denominator.

For the fraction $\frac{1}{3}$, 3 multiplied by 5 will produce 15.

Therefore, the numerator is multiplied by 5 to produce the new numerator:

$$\frac{1\times 5}{3\times 5}=\frac{5}{15}$$

For the fraction $\frac{2}{5}$, multiplying both the numerator and denominator by 3 produces $\frac{6}{15}$. When fractions have like denominators, they are added by adding the numerators and keeping the denominator the same:

$$\frac{5}{15}+\frac{6}{15}=\frac{11}{15}$$

52. B: $90° - 30° = 60°$. Choice *A* is not the correct answer because that is simply the original angle given. Choice *C* is not the correct answer since that is the angle you subtract from. Choice *D* is not the correct answer because that is $90° + 30°$.

53. C: $90°$. To solve, consider the relationships of sine, cosine, and tangent, where θ is a given angle:

$$\frac{\sin\theta}{\cos\theta}=\tan\theta$$

$$\cos\theta = \sin(90° - \theta)$$

Substituting with these relationships and the given angle of 45° we get:

$$x = \tan 45° = \frac{\sin 45°}{\cos 45°}=\frac{\sin 45°}{\sin(90° - 45°)}=1$$

Because $\sin 90° = 1$, the angle for the sine that also equals x must be 90°.

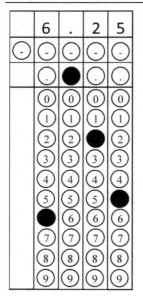

54. If a line is parallel to a side of a triangle and intersects the other two sides of the triangle, it separates the sides into corresponding segments of proportional lengths. To solve, set up a proportion: $\frac{AE}{AD} = \frac{AB}{AC}$, which is $\frac{4}{5} = \frac{5}{x}$. Cross multiplying yields $4x = 25$, or $x = 6.25$.

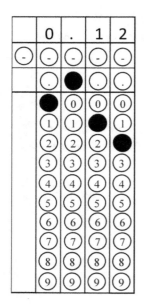

55. The fraction is converted so that the denominator is 100 by multiplying the numerator and denominator by 4, to get $\frac{3}{25} = \frac{12}{100}$. Dividing a number by 100 just moves the decimal point two places to the left, with a result of 0.12.

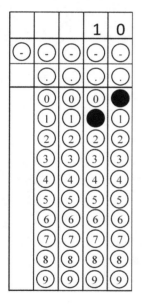

56. 30% is $\frac{3}{10}$. The number itself must be $\frac{10}{3}$ of 6, or:

$$\frac{10}{3} \times 6 = 10 \times 2 = 20$$

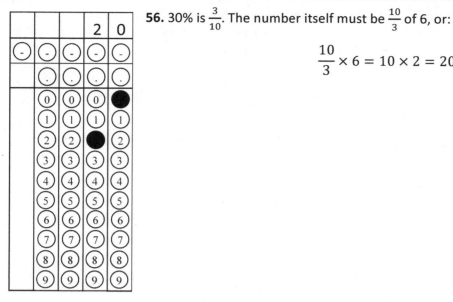

57. 8 squared is 64, and 6 squared is 36. These should be added together to get $64 + 36 = 100$. Then, the last step is to find the square root of 100 which is 10.

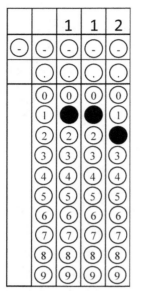

58. Because the 68-degree angle and angle *b* sum to 180 degrees, the measurement of angle *b* is 112 degrees. Because of corresponding angles, angle *b* is equal to angle *f*. Therefore, angle *f* measures 112 degrees.

Dear SAT Test Taker,

We would like to start by thanking you for purchasing this practice test book for your SAT exam. We hope that we exceeded your expectations.

We strive to make our practice questions as similar as possible to what you will encounter on test day. With that being said, if you found something that you feel was not up to your standards, please send us an email and let us know.

We would also like to let you know about other books in our catalog that may interest you.

ACT

amazon.com/dp/1628458844

ACCUPLACER

amazon.com/dp/162845945X

AP Biology

amazon.com/dp/1628456221

CLEP College Composition

amazon.com/dp/1628454199

We have study guides in a wide variety of fields. If the one you are looking for isn't listed above, then try searching for it on Amazon or send us an email.

Thanks Again and Happy Testing!
Product Development Team
info@studyguideteam.com

FREE Test Taking Tips DVD Offer

To help us better serve you, we have developed a Test Taking Tips DVD that we would like to give you for FREE. **This DVD covers world-class test taking tips that you can use to be even more successful when you are taking your test.**

All that we ask is that you email us your feedback about your study guide. Please let us know what you thought about it – whether that is good, bad or indifferent.

To get your **FREE Test Taking Tips DVD**, email freedvd@studyguideteam.com with "FREE DVD" in the subject line and the following information in the body of the email:

   a. The title of your study guide.

   b. Your product rating on a scale of 1-5, with 5 being the highest rating.

   c. Your feedback about the study guide. What did you think of it?

   d. Your full name and shipping address to send your free DVD.

If you have any questions or concerns, please don't hesitate to contact us at freedvd@studyguideteam.com.

Thanks again!

CPSIA information can be obtained
at www.ICGtesting.com
Printed in the USA
LVHW061728020521
686267LV00008B/379